Japan's Withdrawal from International Whaling Regulation

This book examines the impact and implications of Japan's withdrawal from the International Convention for the Regulation of Whaling (ICRW), which came into effect in July 2019.

In 1982 the International Whaling Commission (IWC) adopted a moratorium on commercial whaling which has been in effect ever since, despite the resistance of some countries, first and foremost Japan, Norway and Iceland, that engage in commercial whaling. As one of the key contributors to scientific research and funding, Japan's withdrawal has the potential to have wide-ranging implications and this volume examines the impact of Japan's withdrawal on the IWC itself, on the governance of whaling, and on indigenous and coastal whaling. It provides backgrounds and commentaries on this decision as well as normative and legal discussions on matters relating to sustainable use of resources, and philosophies surrounding whaling in different IWC countries. The consideration of other international environmental regimes, such as the Convention on International Trade in Endangered Species of Wild Fauna and Flora (CITES), is also examined in order to determine the international ripple effect of Japan's decision. The book reveals that this is not just a matter of whaling but one which has significant legal, managerial and cultural implications. Drawing on deep analyses of IWC structures, the book addresses core philosophies underlying the whaling debate and in how far these may influence environmental governance in the future.

This book will be of great interest to students and scholars of environmental law and governance, biodiversity conservation and sustainable development, as well as policymakers involved in international environmental and conservation agreements.

Nikolas Sellheim is an independent consultant on international conservation law with a specialisation on the IWC and CITES. He is also co-Editor-in-Chief of Polar Record in the Scott Polar Research Institute at the University of Cambridge, UK. He has published extensively on different conservation regimes and has conducted two post-docs at Kobe University, Japan, and University of Helsinki, Finland.

Joji Morishita is a retired Professor at Tokyo University of Marine Science and Technology, Japan. Since 2013, he had served as Japan's Commissioner to the IWC and from 2016 to 2018, he served as the Chair of the IWC.

Routledge Studies in Conservation and the Environment

This series includes a wide range of inter-disciplinary approaches to conservation and the environment, integrating perspectives from both social and natural sciences. Topics include, but are not limited to, development, environmental policy and politics, ecosystem change, natural resources (including land, water, oceans and forests), security, wildlife, protected areas, tourism, human-wildlife conflict, agriculture, economics, law and climate change.

Case Studies of Wildlife Ecology and Conservation in India
Edited by Orus Ilyas and Afifullah Khan

Creating Resilient Landscapes in an Era of Climate Change
Global Case Studies and Real-World Solutions
Edited by Amin Rastandeh and Meghann Jarchow

Species, Science and Society
The Role of Systematic Biology
Quentin Wheeler

Conservation Concepts
Rethinking Human-Nature Relationships
Kurt Jax

Japan's Withdrawal from International Whaling Regulation
Edited by Nikolas Sellheim and Joji Morishita

Conservation Leadership
A Practical Guide
Simon Black

Positive Psychology and Biodiversity Conservation
Health, Wellbeing and Pro-Environmental Action
Jolanta Burke, Sean Corrigan, Jimmy O'Keeffe, and Darren Clark

For more information about this series, please visit: www.routledge.com/Routledge-Studies-in-Conservation-and-the-Environment/book-series/RSICE

Japan's Withdrawal from International Whaling Regulation

Edited by Nikolas Sellheim
and Joji Morishita

Routledge
Taylor & Francis Group

LONDON AND NEW YORK

earthscan
from Routledge

First published 2024
by Routledge
4 Park Square, Milton Park, Abingdon, Oxon OX14 4RN

and by Routledge
605 Third Avenue, New York, NY 10158

Routledge is an imprint of the Taylor & Francis Group, an informa business

British Library Cataloguing-in-Publication Data
A catalogue record for this book is available from the British Library

ISBN: 978-1-032-16887-6 (hbk
ISBN: 978-1-032-16889-0 (pbk)
ISBN: 978-1-003-25081-4 (ebk)

DOI: 10.4324/9781003250814

Contents

Figures

Contributors

Steinar Andresen is a Research Professor (emeritus) at the Fridtjof Nansen Institute in Norway. He has published extensively internationally, peer-reviewed articles and books, mostly on international environmental governance. A main analytical focus has been on the effectiveness of these international regimes and how this can be explained. He has been a visiting scholar at the University of Washington, Seattle; Princeton University, Brookings, Washington DC and at the International Institute of Applied Systems Analysis (IIASA). He has also been a professor of political science at the department of political science, university of Oslo.

Gavin Carter is a Founder of Opes Oceani, a global ocean consultancy company that supports businesses, NGOs, government agencies and research bodies working in the oceans sector. Gavin has extensive experience of attending plenary meetings of the IWC and the meetings of other international bodies, such as the Convention on International Trade in Endangered Species of Wild Flora and Fauna (CITES). Gavin was formerly a media adviser and spokesperson in the United States for Japan's Institute of Cetacean Research (ICR).

Ed Couzens is an Associate Professor of Environmental Law in the University of Sydney Law School, Australia. He holds the degrees of LLM Environmental Law (Natal and Nottingham) and PhD (KwaZulu-Natal) and researches and teaches in international law and environmental law. He is the author, co-author, editor or co-editor of more than 70 publications (including books, edited books, book chapters and journal articles). Amongst these are the book Whales and Elephants in International Conservation Law and Policy (Earthscan/Routledge, 2014). Ed has attended six meetings of the IWC between 2007 and 2022, including the five most recent – either as a member of the South African delegation or representing the Asia-Pacific Journal of Environmental Law, for which he is the principal editor. Before joining the University of Sydney in 2015, he taught at the University of KwaZulu-Natal in South Africa for 14 years; and is both an admitted attorney and a field guide in South Africa. His main research focuses are the protection of biodiversity (generally) and wildlife (specifically), especially the movements of migratory species and alien invasive species – at both the national and the international levels.

Malgosia Fitzmaurice is a Professor of International Law at Queen Mary, University of London. She is also a Member of L'Institut de Droit International and Dr *Honoris Causa* of University of Neuchâtel. Malgosia is an Editor-in-Chief of the International Community Law Review and the Queen Mary Studies of International Law. Her research interests are international environmental law, the law of treaties, indigenous peoples' rights and polar law. She taught at many universities, such as University of Berkeley, Sorbonne (Paris 1); University of Kobe, International Foundation for the Law of the Sea (IFLOS). She teaches regularly at the International Law of the Sea Institute of the IMO at Malta.

Cameron Jefferies, S.J.D., is a Professor of Law at the University of Alberta in Edmonton, Canada, where he specializes in international and domestic environmental law, wildlife conservation, energy law and tort law. He completed his graduate degrees at the University of Virginia, School of Law, where he studied as a Fulbright Scholar. Before entering academia, he practiced at Field LLP in Edmonton, Alberta. He has published several book chapters, articles, and books, including Marine Mammal Conservation and the Law of the Sea (Oxford University Press, 2016). He has been an invited speaker at numerous national and international conferences.

Joji Morishita has been Advisor to the Minister of Agriculture, Forestry and Fisheries on International Affairs (Fisheries) since 2019. He has been involved in international oceans and fisheries issues since 1982 as a representative of the Government of Japan, covering bilateral fisheries access and trade negotiations with several countries, meetings of RFMO/As including CCAMLR, CCSBT and NPFC, multilateral fisheries conferences including FAO COFI, APEC Fisheries WG, and also CBD, CITES and UN General Assembly Informal Consultations on the sustainable fisheries resolution, the Meeting on High Seas Fisheries in the Central Arctic Ocean, IUCN Congress, and other international ocean and environmental meetings. Morishita served as Professor at Tokyo University of Marine Science and Technology from 2016 to 2023 and Japan's Commissioner to the IWC from 2013 to 2018. He was the Chair of the IWC from 2016 to 2018 and Chair of the NPFC Scientific Committee from 2015 to 2019. He is currently Japan's Commissioner to the Commission for the Conservation of Antarctic Marine Living Resources (CCAMLR).

David Aarvik Nese holds a Master's degree in Political Science and a Bachelor's degree in International Relations, both from the University of Oslo. For his Master's thesis, he explored the topic of Norwegian whaling, specifically seeking to understand the reasons behind norm non-conformity in his research titled: "Ambiguity, free-riding, or principles? Explaining Norwegian commercial whaling". Nese's research interests lie in statistical analysis and how international norms shape state behavior.

José Truda Palazzo, Jr., a Brazilian environmental consultant, writer and activist whose career spans 45 years, is the founder of the Brazilian Right Whale Research and Conservation Project and has attended IWC meetings for 35 years under diverse capacities as the Brazilian Alternate Commissioner, head of Brazilian delegation to the Scientific Committee, and observer for several NGOs, as well as an adviser to the Uruguayan delegation, and served as CITES Convention delegate for Brazil for several CoPs. Truda Palazzo was also the main proponent of current Brazilian whale watching regulations. A Life Member of the Australian Conservation Foundation and the US National Wildlife Federation, he currently serves as Member of the IUCN Marine Mammals and Protected Areas Task Force and Tourism and Protected Areas Specialist Group, as well as Institutional Development Coordinator for the Brazilian Humpback Whale Institute and as Senior Conservation Officer at IBRACON, the Brazilian Nature Conservation Institute.

Agnes Rydberg is a Lecturer in international law at the University of Sheffield and Deputy Director of the Sheffield Centre for International and European Law. She has extensive experience in the law of treaties, the law of the sea and international environmental law. Agnes also teaches specialist seminars on the law of treaties at the International Maritime Organization. She has previously worked with UN Women and the International Bar Association.

Nikolas Sellheim holds a PhD in law from the University of Lapland, Rovaniemi, Finland. He is the Director of the consulting agency *Sellheim Environmental*. His primary research interests relate to the conservation and utilisation of wildlife and the nexus of human rights for which he has conducted field work in the Canadian commercial seal hunt in Newfoundland as well as in the dolphin-hunting Japanese community of Taiji. Sellheim conducted two postdoctoral research projects at Kobe University (Japan) and the University of Helsinki (Finland). He has been an observer to meetings of the IWC, CITES and other multilateral environmental agreements. Nikolas is co-Editor-in-Chief of Polar Record, the journal of the Scott Polar Research Institute, University of Cambridge.

Akiho Shibata is Professor of International Law and Director, Polar Cooperation Research Centre (PCRC), Kobe University, Japan. He has written in many areas of public international law, including on the ICJ's Whaling case: "ICRW as an Evolving Instrument: Potential Broader Implications of the *Whaling* Judgment," Japanese Yearbook of International Law, Vol. 58 (2015).

Heather Stock is an Associate at Blake, Cassels & Graydon LLP in Calgary, Canada. She holds an BA and JD from the University of Alberta, and she is currently working to complete her LLM degree. Prior to entering practice, Heather completed a clerkship at the Court of King's Bench of Alberta in Edmonton.

Barry Scott Zellen, PhD, is an independent scholar with research interests that include Arctic international relations and the indigenous foundations of world politics. His initial research and writing (1990–2010) focused primarily on the Western Arctic region, where he conducted research during the 1990s while managing several northern, Aboriginal language media organizations in the NWT and Yukon. Since 2010, his research has expanded beyond the Arctic to include indigenous polities of the Pacific basin, from Borneo to Beringia. Zellen has held a Research Scholar appointment in the UConn Department of Geography since 2018, and he was a Fulbright Scholar at the Polar Law Program at the University of Akureyri in 2020. In 2022, he rejoined the editorial team of Intersec: The Journal International Security as its International Arctic Correspondent.

Acronyms

ASW	Aboriginal Subsistence Whaling
ASWSC	Aboriginal Subsistence Whaling Sub-Committee
AWMP	Aboriginal Whaling Management Procedure
AWMS	Aboriginal Whaling Management Scheme
BBNJ	Biodiversity Beyond National Jurisdiction
BWU	Blue Whale Unit
CCAMLR	Convention for the Conservation of Antarctic Marine Living Resources
CCW	(sub-)Committee on Cultures and Whales
CITES	Convention on International Trade in Endangered Species of Wild Fauna and Flora
CMS	Convention for the Conservation of Migratory Species of Wild Animals
COP	Conference of the Parties
DFO	Department for Fisheries and Oceans
EEZ	Exclusive Economic Zone
EIA	Environmental Impact Assessment
FJMC	Fisheries Joint Management Committee
HTA	Aklavik Hunters and Trappers Committee
ICC	International Cetacean Commission
	Inuit Circumpolar Council
ICCPR	International Covenant on Civil and Political Rights
ICJ	International Court of Justice
ICNT	Informal Composite Negotiating Text
ICRW	International Convention for the Regulation of Whaling
IFA	Inuvialuit Final Agreement
IFAW	International Fund for Animal Welfare
IGA	Inuvialuit Game Council
ISA	International Seabed Authority
ISR	Inuvialuit Settlement Region
IWC	International Whaling Commission
JBNQA	James Bay and Northern Québec Agreement
MEA	Multilateral environmental agreement

MMPA	Marine Mammal Protection Act
MSY	Maximum Sustainable Yield
NAMMCO	North Atlantic Marine Mammal Commission
NDP	New Democratic Party
NGO	Non-governmental organisation
NLCA	Nunavut Land Claims Agreement
NMP	New Management Procedure
NMRWB	Nunavik Marine Region Wildlife Board
NWMB	Nunavut Wildlife Management Board
RFMO	Regional Fisheries Management Organisation
RMP	Revised Management Procedure
RMS	Revised Management Scheme
SADC	South African Development Coalition
SAWS	South Atlantic Whale Sanctuary
SC	Scientific Committee (IWC)
	Standing Committee (CITES)
SOF	Japan's Ship and Ocean Foundation
SOWS	Southern Ocean Whale Sanctuary
TAC	Total Allowable Catch
TAH	Total Allowable Harvest
TAT	Total Allowable Take
TRC	Transformative regime change
UNCHE	UN Conference on the Human Environment
UNCLOS	UN Convention on the Law of the Sea
UNDRIP	UN Declaration on the Rights of Indigenous Peoples
UNEP	United Nations Environment Programme
UNHRC	UN Human Rights Committee
WDCS	Whale and Dolphin Conservation Society
WMAC	Wildlife Management Advisory Council
WTO	World Trade Organization
WWF	World Wildlife Fund

Acknowledgments

Nikolas Sellheim My utmost gratitude goes to my wife Gianna Rieb-Sellheim, who does not only lend inspiration and support, but who also provides vivid and constructive challenges to my own thinking, thereby opening up new and untrodden paths of mental directions. Without you, none of this would be possible and words cannot express my appreciation for my best friend, partner and companion. I love you forever and my parts of this book shall be dedicated to you.

The support I have experienced over the last years, which has also contributed to the finalisation of this book, stemmed furthermore from Helene & Eugene Lapointe (merci pour tout!), Meike Hullen & Peter Sellheim, Katharina Sellheim, Yukino Thompson & Konstantin Sellheim, Jessica Rothert & Werner Rieb, my co-editor Prof Joji Morishita, Hideki Moronuki and many more. And of course, the inspirational next generation cannot be forgotten and I'm sure Luka, Elma and Lucca will carry the torch of wisdom into the future. Also the past generation cannot be forgotten, so I furthermore dedicate my parts of this book to my parents. May they rest in peace.

Joji Morishita This book project would not have been possible without my co-editor, Dr Nikolas Sellheim. His knowledge, tenacity, optimism and skill as an editor were instrumental in the creation of this book. In that sense, it is an honor and a pleasure for me to have participated in some small way in the writing and editing of this important and interesting book.

I would like to recognize the Japanese government officials, especially Hideki Moronuki who led the Fisheries Agency team that made possible the unexpected action of Japan's withdrawal from the ICRW. Only time will tell whether the fact that the whaling issue has resulted in Japan's withdrawal from the IWC was good or not. There may be no success or failure in the whaling dispute in the first place. I sincerely hope that this book will provide some insight into this multifaceted and difficult issue.

Finally, my sincere thanks to my wife Nana and my Golden Retriever Lien for providing me with an excuse to relax from time to time amidst the mountain of obligations.

Foreword

Akiho Shibata Sovereign states have the sovereign right not to become a contracting party to a treaty or to withdraw from a treaty in accordance with its provisions. Nothing more, nothing less.

Then, why does Japan's withdrawal from the International Convention for the Regulation of Whaling (ICRW) effective 30 June 2019 attract such a huge attention both academically and politically? This book argues that it is because of its implications, first, for the future of the IWC and its relevance as a global regulatory regime for whales and whaling. Second, and more importantly, the case may have potential profound implications for the future of global environmental diplomacy relating to Earth's biosphere, which is under increasing threat posed by climate change. The lessons that can be learned from Japan's case may be relevant to, as this book suggests, the existing and potential future global regimes that address topics such as conservation of living resources, protection of Earth's ecosystems and biodiversity, utilization of bio/genetic-technology, food security, and, perhaps, animals themselves.[1] These potential far-reaching implications beyond the whaling controversy may be amplified by the fact that the withdrawing nation was Japan who, after the World War, has continuously pursued a diplomacy based on multilateralism and the rule-based international order.

This book aims to address these on-going as well as potential implications of Japan's withdrawal from IWC to "provide for a basis of discussion, both in academia and within the IWC itself" (Introduction). The two co-editors of this book should be acknowledged as the most qualified experts in addressing this complex issue, one from academia with in-depth understanding of international law relating to marine mammals, and the other being a government insider with long-standing experience in the whaling issue both in Japan and internationally. Joji Morishita's personal memoir (Chapter 2) can well be read as an academic reasoning of and the supporting evidence for the official view of Japan's Fisheries Agency, a key actor in Japan's fisheries policy-making and diplomacy on the issues of whaling and marine living resources more generally.

I had the privilege of reading Joji Morishita's book *IWC Withdrawal and International Negotiations*, 2019 in Japanese.[2] In its final chapter, Morishita himself has suggested five lessons for the future of global environmental diplomacy: (1) the role of science, (2) the media and outreach activities, (3) the concept of charismatic animals, (4) the (negative) effect of "perception" and international politics and (5) conflict between globalism and localism. This list clearly suggests the academic complexities requiring multiple and diverse disciplines if one were to address comprehensively the implications of the case and potential solutions for the similar difficulties that the international community would face in the near future. "In an era of multilayered threats, cultural sensitivity and changing environmental paradigms" (Chapter 7), this book is an important first step towards such a grandiose academic endeavor, and the insights provided in each and every chapter contained in this book shall constitute indispensable building blocks for such endeavor

Notes

1 See Anne Peters, *Animals in international law* (Brill 2021).
2 Joji Morishita, *IWC Dattai to Kokusai Kousho* (Seizansha 2009).

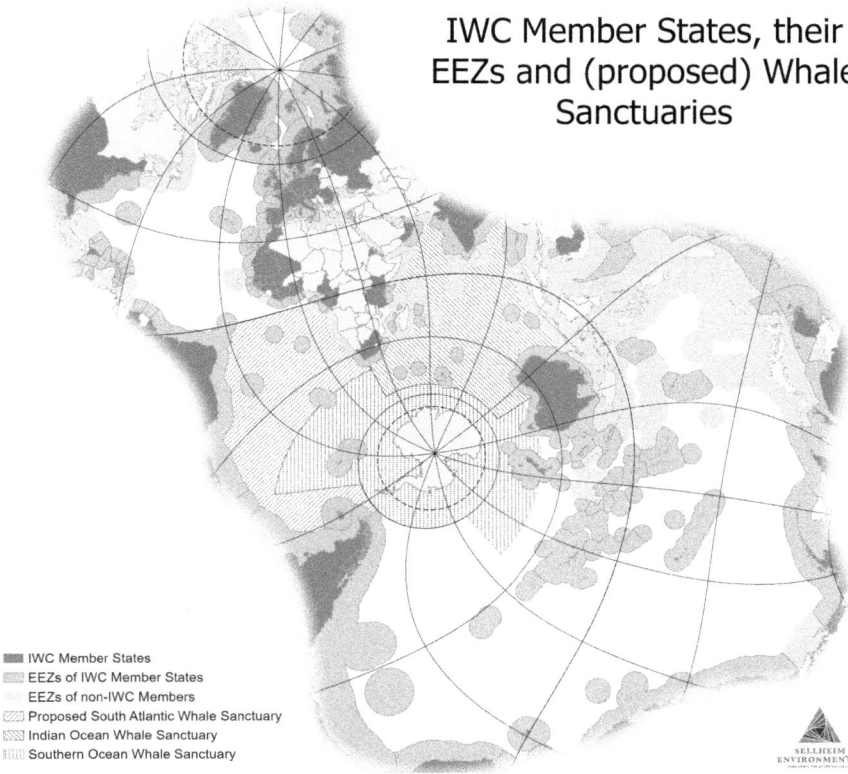

IWC Member States, their EEZs and (proposed) Whale Sanctuaries

IWC Member States
EEZs of IWC Member States
EEZs of non-IWC Members
Proposed South Atlantic Whale Sanctuary
Indian Ocean Whale Sanctuary
Southern Ocean Whale Sanctuary

SELLHEIM
ENVIRONMENTAL

1 Introduction

Nikolas Sellheim and Joji Morishita

Background

On 26 December 2018, Japan officially announced its withdrawal from the 88-member IWC,[1] the management body of the ICRW.[2] This withdrawal took effect on 30 June 2019. While many were taken aback by this decision, the country announced at the 67th meeting of the Commission (IWC67) that it would have to conduct 'a fundamental re-assessment of its position as a member of the IWC',[3] if the different positions concerning commercial whaling could not be reconciled in the Commission. This 're-assessment' was the result of a heated debate on the IWC's move forward: Japan has for more than 30 years pressed for a resumption of commercial whaling and the simultaneous conservation of whale stocks, while the majority of states within the IWC have not agreed to lift the moratorium but instead moved the IWC towards a preservation rather than management authority.[4]

The 30 June 2019 was, therefore, the last day of membership of one of its most important and most controversial members—Japan. After decades of controversy, Japan opted to withdraw from the ICRW and thereby from the IWC. But how did the IWC get there?

The root of the problem is not recent but dates back to the 1970s when the first calls for a ban on commercial whaling, a 'moratorium', were uttered. Already at the 1972 UN Conference on the Human Environment in Stockholm, Sweden ('Stockholm Conference'), the document emerging out of the summit—the Stockholm Declaration and its Action Plan—addressed a moratorium in a very concrete fashion. In Recommendation 33, the Action Plan notes:

> It is recommended that Governments agree to strengthen the International Whaling Commission, to increase international research efforts, and as a matter of urgency to call for an international agreement, under the auspices of the International Whaling Commission and involving all Governments concerned, for a 10-year moratorium on commercial whaling.[5]

While in the course of the 1970s these calls were not followed by the IWC due to the majority of the members either still being actively engaged in whaling or still in the process of phasing out their whaling activities, the end of that decade saw

DOI: 10.4324/9781003250814-1

the joining of Parties with the explicit goal of putting in place a moratorium on commercial whaling. At the meeting of the Commission in 1982, the needed three-fourths majority was finally reached, and the Schedule, the integral element of the Convention setting catch limits and/or geographical limitations, was amended in so far as its Article 10(e) now reads:

> Notwithstanding the other provisions of paragraph 10, catch limits for the killing for commercial purposes of whales from all stocks for the 1986 coastal and the 1985/86 pelagic seasons and thereafter shall be zero. This provision will be kept under review, based upon the best scientific advice, and by 1990 at the latest the Commission will undertake a comprehensive assessment of the effects of this decision on whale stocks and consider modification of this provision and the establishment of other catch limits.[6]

Not surprisingly, not all IWC Members agreed with this zero-catch-quota. Based on Article V.3 of the Convention, disagreeing states have the possibility to lodge objections to any Schedule amendments, provided they do so within a 90-day period. Four IWC Members made use of this provision: Japan, Norway, Peru and the USSR. However, states are also able to withdraw such objection, which Peru did on 22 July 1983 and Japan on 1 May 1987 with regard to commercial pelagic whaling, on 1 October 1987 concerning commercial coastal whaling for minke and Bryde's whales, and 1 April 1988 concerning commercial whaling for sperm whales. For both Peru and Japan, the moratorium is, therefore, binding. Norway and the USSR (now Russian Federation) have not withdrawn their objections and are consequently not bound to it. In other words, they can still engage in commercial whaling, which Norway actively pursues.

The case of Canada and Iceland is different, albeit somewhat similar for the latter.[7] As a sign of disagreement, Canada left the Commission before the moratorium was put into place on 27 June 1981. It was feared that the IWC would also take control over the Arctic narwhal, beluga and right whale hunts, which were conducted by the Inuit. Furthermore, the country feared that its rights under the UN Convention on the Law of the Sea (UNCLOS),[8] which was agreed upon on 10 December 1982 and which contains rights and obligations in the newly established 200 nautical mile Exclusive Economic Zones (EEZs), would be impaired.[9]

Iceland as well took the step of leaving the Commission in 1992, also as a signal of protest against the moratorium and in light of the fact that the North Atlantic Marine Mammal Commission (NAMMCO)[10] was established in the same year. Contrary to the IWC, NAMMCO has never aspired to put a ban on commercial whaling in place. Instead, the organisation promotes the conservation and sustainable use of marine mammals, including whales, irrespective of ethnicity. This is to say, while the IWC put in place the moratorium, it also put in place exemptions for indigenous whalers, known and established as Aboriginal Subsistence Whaling (ASW). Despite the prohibition of all commercial whaling under the moratorium, the ASW regime allows indigenous whalers to hunt whales for subsistence

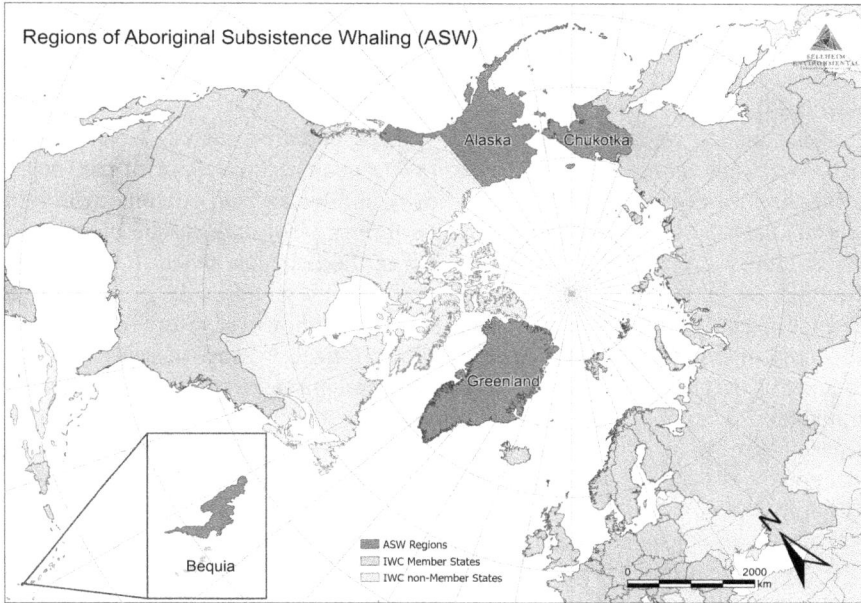

Figure 1.1 Regions of ASW.

purposes, but following a quota set by the Commission. ASW is currently carried out in Greenland, Alaska, Chukotka and Bequia (St Vincent & the Grenadines), as shown in Figure 1.1.

Icelanders, all of whom are considered non-indigenous, were consequently forced to comply with the moratorium since the country failed to lodge an objection. The only possibility to continue whaling was, therefore, to leave the IWC and to join NAMMCO in order to comply with the requirements of the UNCLOS to cooperate through 'appropriate organisations' in the management and conservation of marine mammals. This is not the end of the story, however, and Iceland rejoined the IWC in 2002, yet with a reservation against the moratorium—an issue of contention within the Commission.[11]

Although the moratorium is still in force, originally it was not planned to be in place indefinitely. Instead, as the wording of Article 10(e) of the Schedule reads, it was to be reviewed in 1990. To this end, the Revised Management Procedure (RMP) was developed by the IWC's Scientific Committee. The RMP contained a new methodology for the assessment of whale stocks as well as for their potential utilisation, paying regard to scientific uncertainty and to the role of whales in the marine ecosystem. As such, it was adopted by the Commission in 1994. For the RMP to become operative, it was to be complemented by the so-called Revised Management Scheme (RMS), which was to allow for a scheme for observation and for a system to ensure that quotas were kept. De facto, if the RMS is ever adopted, the moratorium on commercial whaling would be lifted. Since that is the case, the majority of the Commission's members opted for non-adoption, despite the scientific backing that underpins the RMS. Up to this day, therefore, the RMS has not been adopted and at the IWC meeting in 2006, it was officially declared that the IWC had reached an impasse concerning the adoption of the RMS and therewith a potential lifting of the moratorium.

This situation was not acceptable for whaling countries such as Japan, Iceland or Norway, and along with them several other countries in favour of the (potential) lethal use of cetaceans.[12] While Norway and Iceland did continue with their commercial whaling operations due to the former's objection to the moratorium and the latter's reservation to Article 10(e) of the Schedule when it rejoined the Commission, Japan's position was different. Since it withdrew its objection(s) to the moratorium, it was bound to it. This withdrawal, however, was not necessarily made out of free will, but rather because of political pressure, especially by the United States. The US threatened to impose sanctions based on US law which permitted the US to put in place trade barriers if a country undermines an international fisheries agreement, under which the ICRW falls. Since the Damocles sword of the sanctions was dangling over Japan's head because of its objection, it decided to withdraw this objection.[13]

This did obviously not mean the end of the story. Starting from 1987, the Japanese government initiated a large-scale whaling programme—Japan's Whale Research Program in the Antarctic (JARPA)—in the Southern Ocean. According to the Japanese Institute of Cetacean Research, the 'moratorium was introduced due to uncertainties in scientific knowledge about cetacean stocks. JARPA was started to accumulate scientific data and remove this uncertainty'.[14] Contrary to a longstanding narrative,[15] Japan did, therefore, not use a legal 'loophole' to conduct commercial whaling but rather conducted its whale research to show that commercial whaling was indeed possible. Since this also meant the taking of whales in the Southern Ocean and the whale meat was sold on Japanese markets,[16] it was perceived as commercial whaling in disguise.

After Japan's withdrawal from the Convention, its Southern Ocean whaling activities were halted and the country now conducts commercial whaling in its own EEZs. The Total Allowable Catch for 2023 are: minke whale 167, Bryde's whale 187 and sei whale 25.[17] Contrary to the scientific whaling activities in the Southern Ocean, the majority of whales are now being caught along the Japanese coast in six whaling villages, whereas the port of Shimonoseki in south-western Japan constitutes the primary base for commercial whaling operations, as shown in Figure 1.2. By withdrawing from the Convention, Japan, therefore, finally achieved what it strove to achieve within the IWC for decades.

But what does this mean for the IWC? What are the implications of Japan's withdrawal and what could the consequences be? These questions are what this study aims to address from various perspectives. By no means do we aim to have taken into account all relevant issues, nor do we claim to provide a comprehensive picture. After all, we, the Editors, are aware of the highly controversial nature of whaling and do not consider that we have presented 'right' (or 'wrong') positions on the matter. However, the study aims to provide for a basis of discussion, both in academia and within the IWC itself.

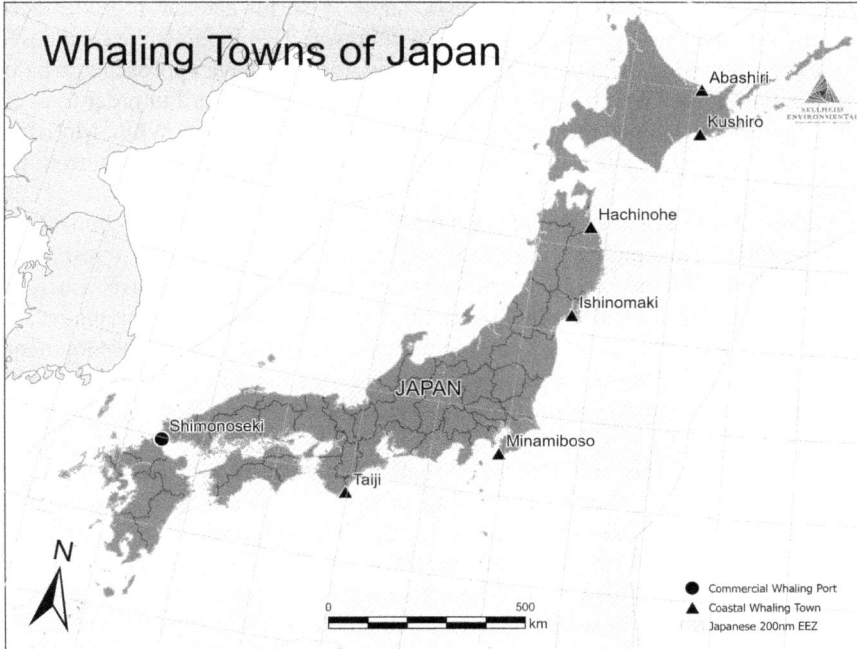

Figure 1.2 Whaling towns of Japan.

The most recent controversy within the IWC circled around the potential establishment of a South Atlantic Whale Sanctuary (SAWS). The SAWS would be the third sanctuary established by the IWC, next to the Indian Ocean Sanctuary and the Southern Ocean Sanctuary.[18] While the SAWS had been proposed on several occasions prior to IWC68 in 2022, what made the meeting special was the fact that when the respective Schedule amendments were taken to a vote, the necessary quorum of 45 (out of now 88) IWC members present for the vote was not reached. It was especially those countries opposing the establishment of the SAWS that were not present in the room. A discussion on what constitutes the quorum as established by the Commission's Rules of Procedure ensued, prompting the IWC to further investigate this issue. Therefore, not only the matter of whether or not to lift the moratorium provides challenges but also procedural aspects, which are related to the way the Commission handles whales and whaling: full-scale protection and further enhancement of protective mechanisms such as in the form of the SAWS, or the possibility for the (lethal) utilisation of whales?[19]

About This Study

In the first chapter, **Joji Morishita** provides his personal views on Japan's withdrawal from the Whaling Convention. Based on his experience as part of the

Japanese delegation and chair of the IWC since 1992, his account provides deep insight into a Japanese perspective on the withdrawal. He pinpoints the problem to the uncompromising manner, countries such as Australia have approached whales and whaling in the IWC, having prevented a proper way forward in order to overcome the deadlock since the putting in place of the moratorium. While Morishita considers the IWC dysfunctional, he also does not see a mass withdrawal from the Convention after Japan has left.

Steinar Andresen and **David Aarvik Nese** open the first sub-section of the study, entitled 'Institutional implications'. The authors tackle the question of Japan's role within the IWC and challenge the IWC in its current form. Since Japan was one of the most contributing members of the organisation. Relying on statistical overviews of the IWC, Andresen and Nese shed light on the developments within the organisation concerning different whaling activities, the development of its anti-whaling stance and Japan's contribution to the different bodies within the IWC.

Cameron Jefferies and **Heather Stock** delve deeply into the protectionist agenda of the IWC by tracing the history of the IWC as turning from a whaling organisation to an organisation aiming for the full protection of cetaceans. The chapter builds on Jefferies' book *Marine Mammal Conservation and the Law of the Sea*[20] in which he called for a compromise agreement between protectionist and utilitarian approaches to whale conservation. However, with Japan's withdrawal from the ICRW, the IWC might now be able to further advance its protectionist agenda by focusing on significantly more important threats to whales than whaling: climate change, by-catch or ship strikes.

In Chapter 6, **Nikolas Sellheim** considers whether Japan's withdrawal from the ICRW might have implications for other international treaties. This is particularly relevant for the Convention on International Trade in Endangered Species of Wild Fauna and Flora, which has been facing similar challenges as the IWC. By focusing on countries from the Southern African Development Community, he argues that while it is not likely that ripple effects will occur, other modes of environmental governance in light of the multilayered threats to flora and fauna are necessary.

Part II of the volume concerns 'Cultural considerations' that **Malgosia Fitzmaurice** and **Agnes Rydberg** address by looking at indigenous whaling after Japan's withdrawal. The authors argue that commercial whaling will no longer play a significant role in the Commission and that the subject of controversy will likely be surrounding ASW. While ASW is largely accepted, calls for increasing quotas, the ambiguous definition of ASW itself and the killing methods, which are comparably more cruel than killing methods in commercial whaling, will shape the debate.

Indigenous whaling is also the topic of **Barry Scott Zellen's** chapter. However, he focuses on the processes and developments that led to Canada's withdrawal from the Whaling Convention in 1982 over concerns over the impacts of the whaling moratorium on Inuit bowhead whaling in the Canadian Arctic. In light of Japan's exit from the IWC, he addresses the question of whether lessons for Japan's reinstatement of coastal whaling can be learnt from Canada's exit and the effects on Inuit bowhead whaling.

In the last chapter of this part, **Nikolas Sellheim** looks at the way whales and whaling are perceived in the different cultures of IWC member states. Taking Durkheim's dichotomy of the 'sacred' and 'profane' into account, he shows how whales and whaling have taken culturally different forms. Given this major gap amongst IWC member states, he proposed the establishment of a Committee on Cultures and Whales in order to overcome culturally imposed gaps and to find a way of cultural communication on matters concerning whales and whaling.

The final part of the study, 'Perspectives', contains two personal commentaries by long-term practitioners on matters of the IWC. On the one hand, **Gavin Carter** addresses the commercial impacts of Japan's withdrawal from the Whaling Convention. In his view, Japan's withdrawal may raise questions concerning the role of international treaties and the obligations for the protection of ecosystems especially with regard to the way science is used for decision-making. While that may be so, however, Carter concludes that the commercial impacts are minimal since the benefits of cooperation still prevail.

The last chapter by **José Truda Palazzo, Jr.** provides indeed an epilogue to the 'whaling wars'. Palazzo considers Japan's withdrawal as a beneficial event to the IWC since this enables countries that have an interest in the non-lethal use of whales to further their agenda within the Commission. This is especially the case for IWC members of the southern hemisphere who are now able to protect whales significantly better, given that Japan has now abandoned its whaling operations in the Southern Ocean. Also for Japan, he argues, the withdrawal will be beneficial: after all, whaling will no longer be an issue that affects foreign policy, enabling the country to become a leading marine conservation partner.

We, the Editors, have decided not to include a concluding chapter in the study. The reason is rather simple: we do not consider the matter closed, in so far as the issues that are addressed within this study constitute merely a snapshot of topics that are of relevance. The study should serve as a start of a discussion and not as a concluding end. With this study, we, therefore, hope to provide some views on the benefits and disadvantages of Japan's exit from the IWC and wish for more fruitful dialogue that contributes to effective and equitable decision-making concerning whales and whaling.

Notes

1 The members of the IWC at the time of writing are: Antigua & Barbuda, Argentina, Australia, Austria, Belgium, Belize, Benin, Brazil, Bulgaria, Cambodia, Cameroon, Chile, China, Colombia, Republic of the Congo, Costa Rica, Côte d'Ivoire, Croatia, Cyprus, Czech Republic, Denmark, Dominica, Dominican Republic, Ecuador, Eritrea, Estonia, Finland, France, Gabon, The Gambia, Germany, Ghana, Grenada, Guinea-Bissau, Hungary, Iceland, India, Ireland, Israel, Italy, Kenya, Kiribati, Republic of Korea, Laos, Liberia, Lithuania, Luxembourg, Mali, Marshall Islands, Mauritania, Mexico, Monaco, Mongolia, Morocco, Nauru, Netherlands, New Zealand, Nicaragua, Norway, Oman, Palau, Panama, Peru, Poland, Portugal, Romania, Russian Federation, San Marino, Sao Tome & Principe, St Kitts & Nevis, Saint Lucia, St Vincent & The Grenadines, Senegal, Slovak Republic, Slovenia, Solomon Islands, South Africa, Spain, Suriname, Sweden, Switzerland, Tanzania, Togo, Tuvalu, UK, Uruguay, USA.

2 International Convention for the Regulation of Whaling, 2 December 1946, 161 UNTS 72.
3 Masaki Taniai, Intervention at the International Whaling Commission, 13 September 2018. Audio on file with author.
4 Robert L Friedheim (ed.), *Toward a sustainable whaling regime* (UWP 2001).
5 United Nations, Report on the United Nations Conference on the Human Environment. 1972, A/CONF.48/14/Rev1.
6 ICRW, Schedule, Article 10(e).
7 See Couzens' chapter in this book.
8 United Nations Convention on the Law of the Sea, 1833 UNTS 3, 10 December 1982, in force 16 November 1994.
9 Nikolas Sellheim, *International Marine Mammal Law* (Springer 2020).
10 NAMMCO comprises four Arctic countries: Iceland, Norway, Greenland and the Faroe Islands.
11 See Couzens' chapter in this volume.
12 Especially Antigua & Barbuda, St Lucia, Guinea and several other developing states are part of those countries vehemently working for the lifting of the moratorium. In the past, this has led to allegations of vote-buying by Japan in order to achieve the necessary numbers to lift the moratorium. See Jonathan R Strand and John P Tuman, 'Foreign aid and voting behavior in an International Organization: The case of Japan and the International Whaling Commission', (2012), *Foreign Policy Analysis* 8(4), 409.
13 *New York Times*, 'Japan agrees to end whaling' *New York Times* (New York, 6 April 1985) 1.
14 "このモラトリアムは、鯨類資源に関する科学的知見の不確実性を理由に導入されました。 JARPAは、科学的データを蓄積し、この不確実性を取り除くために開始されました" (ICR, '南極海鯨類捕獲調査(JARPA)の概要'. < https://www.icrwhale.org/JARPAgaiyou.html> accessed 20 April 2023.
15 Jeffrey D Lindemann, 'The dilemma of the International Whaling Commission: The loophole provisions of the commission vs. the world conscience', (1998), DJILP 7, 491.
16 All in line with the Convention, which reads in Article VIII.2 on 'special permit' whaling (i.e. scientific whaling): "Any whales taken under these special permits shall so far as practicable be processed and the proceeds shall be dealt with in accordance with directions issued by the Government by which the permit was granted."
17 Ministry of Foreign Affairs Japan, 'Japan and the management of whales'. https://www.mofa.go.jp/policy/economy/fishery/whales/japan.html accessed 20 April 2023.
18 For a map of the proposed and existing sanctuaries, please refer to the opening map of this volume.
19 On a summary of the issue related to the quorum at IWC68, see Sellheim Environmental, 'The blurry question of the quorum', (2022), TCLD 1(4),10.
20 Cameron S.G. Jefferies, *Marine Mammal Conservation and the Law of the Sea* (OUP 2016).

2 A Memoir – Japan's Road to Withdrawal from the International Whaling Commission

Joji Morishita

Introduction

On 26 December 2018, Japan announced its decision to withdraw from the International Convention for the Regulation of Whaling (ICRW) that established the IWC. While the announcement was a surprise for some, for others it was an inevitable outcome of the international controversy surrounding whales and whaling.

The author had been involved in the whaling issue since 1992 when he attended his first IWC meeting and served as Japan's Commissioner to the IWC from 2013 until its withdrawal. He also chaired the 67th IWC meeting held in September 2018 at Florianópolis, Brazil. The results of the 67th IWC meeting had led Japan to the withdrawal. However, the withdrawal was not an "on-the-spot" decision, but changes in Japan's negotiation approach, which eventually resulted in the withdrawal, had begun from around 2014; changes that could be characterized as Japan's last attempt for resolving the whaling controversy in the IWC framework. The author recollects the proposals and discussions from the 65th IWC meeting in 2014 to the 67th meeting in 2018 mainly from Japan's viewpoint and elaborates why the attempt, like other previous attempts by several IWC Chairs for "peace making," eventually failed, hoping that it will present some important lessons for other international controversy arising from fundamental differences of positions among stakeholders.

This chapter will discuss the background and reasons behind Japan's decision to withdraw from the international organization, as Japan has always taken a cooperative approach within the international community; what will happen to the IWC as a result of Japan's withdrawal; and what impact is expected for the future of the conservation and management of fishery resources and biodiversity. He will also touch on the Antarctic whaling case before the International Court of Justice (ICJ), which was decided in 2014.

The author has attended approximately 20 IWC Plenary Meetings and many related meetings as a member of the Japanese delegation since 1992 and served as the Japanese Commissioner to the IWC from 2013 to 2019 and as the IWC Chair from 2016 to 2018, so he has had access to information that is not publicly available. However, this chapter is written based on the information publicly available.

DOI: 10.4324/9781003250814-2

Furthermore, the views and interpretations expressed in this chapter are the author's own and do not represent the position of the Government of Japan.

Shifts in the Issues Concerning Whaling

The Issue of Whaling Has Shifted over Time

In response to the past overhunting, the IWC banned the take of humpback and other whales in the Antarctic Ocean beginning in the 1960s and introduced resource conservation and management measures known as the New Management Procedure in the 1970s. However, concerns were expressed that the measures were insufficient to deal with various scientific uncertainties with the scientific level at the time. This led to the adoption of the moratorium on commercial whaling in 1982. In other words, the main issue at that time was the science surrounding whale stock assessment and management. In fact, Paragraph 10(e) of the ICRW Schedule, which defines the moratorium on commercial whaling, provides as follows:

> *(e)* Notwithstanding the other provisions of paragraph 10, catch limits for the killing for commercial purposes of whales from all stocks for the 1986 coastal and the 1985/86 pelagic seasons and thereafter shall be zero. This provision will be kept under review, based upon the best scientific advice, and by 1990 at the latest the Commission will undertake a comprehensive assessment of the effects of this decision on whale stocks and consider modification of this provision and the establishment of other catch limits.[1]

This is a procedure to resume commercial whaling, so to speak, by temporarily suspending commercial whaling, during which time a comprehensive scientific stock assessment will be conducted, and a non-zero quota will be considered. In response to this provision, the IWC Scientific Committee developed the Revised Management Procedure (RMP) in 1992 and completed a catch limit calculation method that would allow the sustainable use of whale stocks without depletion, even under scientific uncertainty. Dr. Philip Hammond, then Chairman of the Scientific Committee, declared:

> Thus, one of the most interesting and potentially far-reaching chapters in the science of natural resource management came to a conclusion. The Commission could now put in place a mechanism for the safe management of commercial whaling, regardless of whether or not the 'moratorium' was lifted.[2]

However, the completion of the RMP did not solve the whaling problem. The anti-whaling countries argued that strict monitoring and enforcement measures were necessary to ensure compliance with the safe catch limits calculated by the RMP and proposed the completion of the Revised Management Scheme (RMS) incorporating these monitoring and enforcement measures as a new condition for resuming commercial whaling. The debate shifted from science to monitoring and enforcement. After nearly 15 years and 50 meetings, the discussions on the RMS finally came to a halt. As discussions on the RMS progressed, anti-whaling nations such as Australia began to argue that the completion of the RMS did not mean the resumption of whaling. One of the reasons for this is the issue of commerciality.

The commercial whaling moratorium was originally intended to address scientific uncertainties in whale resource management and to consider a non-zero catch limit "by 1990 at the latest."[3] Over time, however, the image of the commercial whaling moratorium as a permanent ban on commercial whaling as an illegal activity grew. Furthermore, the ban on commercial whaling also led to the image that commerciality was a bad thing. Considering that the reason for the moratorium on commercial whaling was to deal with scientific uncertainty in resource management, and not the existence of commerciality, this is a misrepresentation of the issue. However, under the reality that many anti-whaling countries in the IWC are opposed to commerciality, a catch limit for whaling cannot be expected to be established. Of course, Japan had made proposals eliminating commerciality as much as possible, but it is always possible to argue that there is still a certain amount of commerciality in the proposed whaling under the modern monetary economy. No clear reason has been given as to what is wrong with commerciality in the first place, and most human activities have a commercial nature. Business, the pursuit of profit, is a legitimate activity.

The shift in the debate on the whaling issue continues. In recent years, there has been a growing belief that whales are special, a so-called species of "charismatic megafauna" that should not be allowed to be taken under any circumstances. This charismatic megafauna is said to include elephants, tigers, wolves, sharks, etc. and is a further step toward the concept of animal rights, which demands the same rights for all animals as for humans, and the concept of animal welfare. Under the concept of charismatic megafauna, all animals are not equal. Here, whaling is an unacceptable activity under any circumstances.

The debate over whaling is often described as emotional and irreconcilable. Initially, at least on the surface, there were scientific and legal arguments, and there was hope that once these were settled, whaling would be allowed. Now, however, the debate over whaling has become a fundamental difference in thinking about whales and whaling, a conflict between the view that whales, like other marine living resources, can be used in a sustainable manner based on scientific information and the view that they are charismatic animals that should be absolutely protected.

Repeated Failure of "Peace Negotiations"

Since the 1990s, the IWC has repeatedly held "peace negotiations" to resolve international conflicts over whaling, in which compromises were sought between the anti-whaling nations and the supporters of sustainable use of whales (the pro-whaling nations). The four chairpersons of the IWC have led these "peace negotiations" because they felt threatened by the situation in which confrontations prevailed, and emotionally provocative discussions, which were uncharacteristic of an international organization, were rampant.

The first of these negotiations was proposed by Ireland in 1997, when at the 49th Session of the IWC (Monaco), Michael Carney (Ireland), who later became the IWC Chair, expressed his concern that the IWC was at risk of collapsing because of the stalemate over whaling and that whaling activities other than Aboriginal Subsistence Whaling (ASW) were being conducted outside the control of the IWC. In order to overcome this situation, Ireland proposed a compromise package that included: (1) the catch limits should be established only for existing coastal whaling, with a worldwide ban on whaling in other areas; (2) whale products should be for local consumption only, with international trade prohibited; and (3) the issuance of special scientific permits (research whaling) under Article 8 of the ICRW should be phased out. Many countries, including Japan, appreciated the Irish proposal and expressed readiness to consider it further, but the proposal failed, partly because of the attitude of anti-whaling nations that they would not accept any whaling other than ASW. For the anti-whaling nations, the proposal included a number of advantages, such as a ban on whaling in the high seas, a phase-out of research whaling, and a ban on international trade in whale meat and other products. However, no compromise was reached because the package included the resumption of coastal commercial whaling. For the anti-whaling nations, the only way forward would be to ban all commercial whaling.

The negotiations to develop the RMS did not bury the fundamental differences between the anti-whaling and pro-sustainable use countries' positions on whales and whaling but rather appeared to sharpen them. Both sides needed to cede something and work out a compromise in order to achieve an agreement, but the mutual distrust was so great that some of the anti-whaling nations, taking a hard-line position, even went so far as to express a position that undermined the very foundation of the RMS negotiations, arguing that the completion of the RMS did not mean acceptance of commercial whaling. Concerned about this situation, Henrik Fischer (Denmark), the Chair of the IWC, established an RMS sub-group with representatives of a limited number of countries and, based on the discussions in this group, submitted a proposal for an RMS package to the 56th session of the IWC in Sorrento (Italy) in 2004. However, this proposal also received a chilly response, especially from anti-whaling nations, and was effectively suspended at the 58th Session of the IWC in St. Kitts and Nevis in June 2006.

Another "peace negotiation" is the "Future of the IWC" project, which was handed over from Chair Bill Hogarth (U.S.) to Christian Maquiera (Chile). At the 59th Annual Meeting in Anchorage in 2007, the hard-line anti-whaling nations remained insistent on protecting whales, and no progress was made in resolving the issues rooted in the fundamental differences in positions over whales and whaling. In addition, Japan made a statement on the last day of the meeting that it would fundamentally review its relationship with the IWC, following the rejection of Japan's request for a small-type coastal whaling catch limit, which included all kinds of concessions, at this meeting where the extension of the catch limits for the ASW was approved, highlighting the critical situation of the IWC.

The outcome of the Anchorage meeting, including Japan's statement, caused a great deal of stir and concern, even among anti-whaling officials. In particular, Chair Hogarth (U.S.), who was concerned that the possibility of the IWC's collapse had become a reality, proposed the "Future of the IWC" project. The project conducted meetings and consultations vigorously, but at the 61st Plenary Meeting in Madeira (Portugal) in June 2009, the activities of the project were extended for one year on the grounds that the discussions were not yet complete. In February 2010, IWC Chair Maquiera (Chile), with a vision to improve both the conservation and management of whale stocks, presented the idea of allowing whaling activities during the next 10-year interim period by removing the whaling categories of commercial whaling, research whaling, and ASW, but at the cost of a reduced scale of take from the current level. This elimination of the whaling category was an attempt to avoid any opposition based on the rigid government policy of anti-whaling member countries and to increase the possibility of compromise by not specifying the category of whaling. Since the overall number of whales to be taken would be reduced from the current scale, this was not a bad deal for the anti-whaling nations.

However, Australia, the most hard-line anti-whaling nation, immediately responded to the Maquiera proposal by announcing a counter-proposal that included the reduction and abolition of whaling activities in the Antarctic Ocean within five years. Australia's position is that it will not accept the agreement unless commercial whaling and research whaling are abolished with a clear deadline. Some of the so-called anti-whaling nations were open to a compromise that would allow some whaling activities, but as long as there were hard-line anti-whaling nations that would not allow any whaling, it was impossible to reach an agreement.

On 22 April 2010, the "Proposal by the Chair and Vice-Chair of the IWC on the Future of the IWC," which included specific numbers of whales to be taken over a 10-year period, was submitted as a compromise based on the Maquiera proposal. On 31 May, however, the Australian government filed a lawsuit in the ICJ seeking to halt Japan's research whaling program in the Antarctic Ocean. This filing effectively meant the end of the "Future of the IWC" project. By resorting to litigation rather than consensus building to resolve the issue of research whaling, which was key to a comprehensive compromise agreement, Australia has abandoned dialogue within the IWC.

The 62nd IWC session failed to reach a consensus decision on the "Future of the IWC" project due to the wide divergence in basic positions on whales and whaling among the member nations. Japan requested that efforts be made to achieve a consensus decision in accordance with the proposal of the Chair and Vice-Chair, but Australia, Latin American countries, and other hard-line anti-whaling nations refused to discuss the issue on the basis of the proposal. This was the collapse of the "Future of the IWC" project as the last "peace negotiation."

Thus, all of the IWC's "peace negotiations" have failed. Why? On the one hand, there are countries that want to use whales sustainably as a marine living resource, and on the other hand, there are countries that do not want to allow the taking of even one whale because they believe that whaling to kill whales, which are special (charismatic) animals, is evil to begin with. In normal international negotiations, a middle ground is sought between the two sides and some kind of compromise is worked out. However, in the case of the IWC, the middle ground has led to the breakdown of negotiations because of the acceptance of certain whaling activities, which are inevitably included in the middle ground. From the perspective of the anti-whaling nations, negotiating and allowing even a small catch limit of whaling is akin to negotiating with terrorists and allowing their activities, which are essentially evil. We do not negotiate with terrorists. If they become engaged in negotiations with pro-sustainable use nations such as Japan and accept even a small catch limit, the governments would face strong criticisms from the general public and non-governmental organizations (NGOs). The history of "peace negotiations" at the IWC over the past 30 years has clearly shown the structure and character of the whaling issue, where no negotiation or dialogue has been possible.

The Antarctic Whaling Case at the ICJ

In 2010, Australia filed a complaint with the ICJ on the grounds that Japan's Whale Research Program in the Antarctic Phase II (JARPAII) was a quasi-commercial whaling program and therefore in violation of the moratorium on commercial whaling, etc. After more than four years of deliberations, on 31 March 2014, the ICJ ruled that JARPAII did not fall within the scope of the scientific research under a special permit as stipulated in Article 8.1 of the ICRW, and as a result, ordered the cancellation of JARPAII. Since there have been numerous legal analyses of the content of this ruling, this chapter will not repeat the analysis and instead consider the relationship between the conflict at the IWC and the lawsuit at the ICJ.[4]

Australia filed its lawsuit with the ICJ on 31 May 2010. The U.S. and New Zealand, which had played a leading role in promoting the "Future of the IWC" project, are said to have urged Australia not to pursue the lawsuit, because the lawsuit practically meant denial and collapse of the project for building compromise among the IWC members. But why was Australia not persuaded?

The Prime Minister of Australia in 2010 was Kevin Rudd of the Labor Party, who was said to be pro-China. He became the Prime Minister in December 2007 and his first trip abroad as Prime Minister in April 2008 was to China, but the trip did not include Japan. That became a big news at that time in Japan. He was very interested in environmental issues, and in the November 2007 parliamentary election in which he won the prime minister's seat, he included the whaling issue as one of the campaign issues and mentioned a possibility of ICJ litigation.[5] He also appointed rock singer and hard-line anti-whaling advocate Peter Garrett as the Minister of the Environment. Prime Minister Rudd maintained a high approval rating for about two years after taking office, but his approval rating declined sharply in late 2009, and he did not run in the party leadership election on 24 June 2010, immediately after the ICJ complaint. He was unseated by Deputy Prime Minister Julia Gillard without vote. Therefore, most interpretations are that the ICJ complaint was one of the means to regain domestic support in Australia.[6]

While New Zealand later joined the ICJ lawsuits, it seemed that its domestic opinion to join the lawsuits was raised because of Australia's ICJ action. My impression at the ICJ proceedings was that New Zealand remained reserved throughout the process.

Australia had chosen to take the whaling issue to court, but I would like to compare the implications of taking the resolution of this issue to the debate in the courts with those of trying to resolve the issue through the IWC discussions.

First, it should be noted that the ICJ Whaling Case only covered the JARPAII in the Antarctic Ocean. It did not cover Japan's other whale research program in the North Western Pacific Ocean, the JARPNII. In other words, the ICJ judgment[7] does not cover the entire whale research programs, much less the entire whaling activities. Nor is the issue of whaling only about whether or not to conduct research whaling. The whaling issue is a group of many diverse problems, including a proposal to establish a South Atlantic Whale Sanctuary, catch limit proposals for Japan's small-type coastal whaling, interpretation of the purpose of the ICRW, animal welfare issues regarding whales, and ASW. Antarctic research whaling, which was the subject of the ICJ whaling case, is certainly an important element of the whaling problem, but its resolution does not mean that the whole whaling problem has been resolved.

Furthermore, the ICJ ruling is only binding on the parties to the whaling case. Prior to Japan's withdrawal from the IWC, 89 countries were members of the IWC, but only 3 of them were parties to the ICJ ruling. Precisely, 86 countries are at least legally unaffected by the ICJ ruling and can ignore the ruling. So how did the IWC as a whole react to the ICJ ruling? To put it simply, the anti-whaling nations, including Australia, which are parties to the ICJ ruling, did not use the ruling as a basis for discussion at the IWC, at least not actively. In this author's view, it was even disappointing.

Immediately after the ICJ ruling, many NGOs and the media expected Australia and New Zealand to make a high-profile declaration of victory at the 65th IWC session in Slovenia in September 2014 in response to the ICJ ruling, calling on the IWC to ban whaling completely. Indeed, a draft resolution was submitted by New Zealand that included procedures to make it more difficult to consider and implement research whaling programs, and this was adopted. Both countries welcomed the ICJ ruling at the session. However, instead of going deeper, they shifted the focus to the discussion for the adoption of the resolution, and the IWC as a whole considered the draft resolution with little discussion on the ICJ ruling itself.

Why? For one thing, although the ICJ ruling declared JARPAII illegal and called for its suspension, it did not deny the existence of the research whaling program itself as defined by the ICRW and even stated that it expected Japan to consider the ruling when planning new whale research programs in the future and did not make a judgment on whether the research whaling program itself was acceptable or not. In other issues of the case, the court made judgments that did accept Japan's arguments. For example, Australia argued that the purpose of the ICRW had changed over time and that the main purpose of the Convention now was to protect whales. The ICJ concluded, however, that the purpose of the Convention remained unchanged and includes the sustainable use of whale resources. The ruling did not deny lethal research to take whales and recognized cases where it is reasonable in light of the purpose of the research. The ICJ also pointed out that the ICRW permits the sale of whale meat after research whaling, which anti-whaling groups cite as a basis for characterizing Japan's research whaling as a pseudo-commercial whaling.[8] These have been points of contention in discussions at the IWC, and the ICJ's decision was basically in line with Japan's arguments.

In joining the 65th IWC session, there was one common understanding among Japan, Australia, and New Zealand. That is, there was no merit in, and it would not be productive in repeating the ICJ's arguments or discussing differing interpretations of the ICJ decision at the IWC session. Also, other IWC members did not seek to discuss the judgment in detail.

Japan's Response to the ICJ Ruling

In response to the ICJ ruling, on 31 March 2014, the Japanese government issued a statement by Chief Cabinet Secretary Suga, stating that "Japan is disappointed and regrets that the Court ruled that JARPAII by Japan did not fall within Article VIII, paragraph 1, of the ICRW. However, Japan will abide by the Judgment of the Court as a State that places a great importance on the international legal order and the rule of law as a basis of the international community."[9] On 18 April, Minister Hayashi of Agriculture, Forestry, and Fisheries issued a statement in which he indicated Japan's policy on how to respond to the ICJ ruling. Specifically, he stated that "Japan has confirmed its basic policy of pursuing the resumption of commercial whaling, by conducting research whaling, through the cooperation among the ministries concerned, based upon international law and scientific evidence in order to gather scientific data that is essential for the management of whale resources" and that Japan would suspend JARPAII, which is the subject of the ruling, and would develop and implement a new Antarctic whale research program to meet the various conditions called for by the ruling for future whale research programs.[10]

Japan's response is based on the fact that the ICJ ruling recognized the sustainable use of whales as a resource as one of the objectives of the ICRW, and that the ruling did not reject the research whaling program itself but rather assumed that Japan could conduct new research whaling programs in the future. While the anti-whaling nations may not agree with this response, it is also a response in line with the ICJ ruling.

So how did the ICJ ruling affect Japan's whaling policy? In particular, what does the ruling have to do with Japan's withdrawal from the IWC? In other words, would Japan have withdrawn from the IWC without the ICJ ruling?

The author's feeling is that, even without the ICJ ruling, as the approach of attempting to resolve the whaling issue through IWC negotiations had already broken down, Japan would have to choose the option of withdrawing from the Convention. In fact, it was before March 2014, when the ICJ ruling was issued that Japan changed its negotiating policy, assuming that it could eventually lead to withdraw from the Convention.

Change in Japan's Negotiating Policy

After years of negotiations, it has become clear that the anti-whaling nations will not accept whaling even if it is scientifically possible to use whale resources in a sustainable manner without depleting them, even if we introduce monitoring and control measures to ensure compliance with catch limits and other regulations, and even if we accept a drastic reduction in the scale of whaling. A three-fourths vote is required at the IWC to resume whaling by voting. In recent years, the "Sustainable Use group," which supports whaling, has had around 40 votes and the anti-whaling group has had almost 50 votes in the IWC, and given the solidarity of the anti-whaling EU group and the South American countries known as the "Buenos Aires Group," the Sustainable Use group must secure around 100 more votes as a new member country to support sustainable use to reach three-fourths. This will be extremely difficult to achieve.

Based on this recognition, Japan shifted its negotiating policy from the 65th session of the IWC in 2014. Continuing to discuss science, regulatory measures, and compromises is like studying math to pass an English exam; there is a mismatch between the issues to be discussed and the issues to be resolved. What then should be discussed and what should be the goal? The response to this was a shift in negotiating policy. In other words, the issue to be discussed is the essential issue of whether the IWC can and will work to avoid conflict, achieve the objectives of both the conservation and management of whale stocks, and restore the function of the IWC as an international organization, on the premise that there are two groups in the IWC with fundamentally different positions regarding whales and whaling.

In order to clarify that the issue was not science or regulatory measures but a fundamental difference in position regarding whales and whaling, Japan asked specific questions repeatedly at the 65th IWC session, such as where in the moratorium clause on commercial whaling a permanent ban on whaling is stipulated, and what the additional scientific information that is needed to change the position against whaling would be. The results were, as expected, no specific rebuttals and answers to the questions, but of the sort that said they were opposed to whaling and that the moratorium should not be repealed. In response, Japan proposed to discuss the above fundamental issues at the 66th IWC session in 2016. The idea was to study English in order to pass the English exam. The anti-whaling countries supported this proposal, saying that any dialogue was welcome, but no concrete solutions were offered, and the traditional opinion to the effect that they oppose whaling because they are against it was repeated. Some anti-whaling countries did not even accept this essential discussion in public. This is because even exchanging views or negotiating with the supporters of sustainable use of whaling, which is an absolute evil from their perspective, would be subject to public and political criticism.

And at the 67th IWC session in 2018, Japan submitted a proposal aimed at agreeing to disagree, therefore, not pursuing compromises, and coexisting within the IWC without getting in each other's way. For Japan, this is a major concession, but it is also the last attempt to achieve the resumption of whaling in the

IWC forum. However, this proposal was also rejected in the vote, and on the other hand, the Florianópolis Declaration, proposed by Brazil, was adopted to the effect that the IWC should evolve as a whale protection organization. Even coexistence of different positions in the IWC was denied. The Japanese delegation therefore declared that it had no choice but to review its relationship with the IWC, and after serious and careful consideration after returning to Japan, it came to the decision to withdraw from the IWC.

Japan's Policy Goals in the Whaling Issue: Two Pillars

What are Japan's policy objectives in the whaling issue? If this is seen only as the resumption of whaling and the preservation of whaling culture through it, the essence of the whaling issue will be missed. It will only see one aspect of the whaling issue and will fail to understand Japan's policy or will make misguided criticisms and assumptions.

Japan's whaling policy has two major pillars. One is the resumption of commercial whaling, which was interrupted by the introduction of a moratorium on commercial whaling in 1982. This policy goal was addressed through withdrawal from the IWC. What, then, is the other policy goal?

When people are interviewed on street corners about the whaling issue, they often make comments to the effect that "it is outrageous that anti-whaling countries oppose eating whale meat while they eat cows and pigs." This may seem to be a simplistic opinion at first glance, but it raises many important issues. For example, who decides what is good to eat and what is not good to eat? If someone else decides, do we have to follow that decision? Can we force others to accept that decision? What is the basis for the argument that livestock can be eaten but wildlife should be protected? These many issues encompass a wide variety of arguments and relate to and affect a wide range of areas far beyond the whaling issue. The author cannot find the right words to describe them succinctly and comprehensively, but this is the other policy goal that is contained in the whaling issue.[11]

Once again, these issues can be expressed, for example, from the perspective of the whaling issue as follows: Is it acceptable that the values of anti-whaling nations, which do not regard whales as a resource, despite the fact that it is scientifically and legally possible to use whales as a resource in a sustainable manner, should be forced on nations, including Japan, which regard whales as a resource that can be used in a sustainable manner, in the form of decisions by the international organization, the IWC? Can the international community decide on values through a vote and then force them onto the losing side? Contrary to the created image that only a small number of IWC member countries support whaling, about 40 countries are supporters of sustainable use in the IWC, including many developing countries in addition to whaling nations such as Norway, Iceland, and Russia. Under this condition, can the IWC force anti-whaling decisions adopted by vote?

Their interest and reason for joining the IWC is the "other policy goal" mentioned above. In their statements at the IWC, we often hear assertions to the effect that

the use of living marine resources must be based on scientific evidence, that the imposition of values is environmental imperialism and environmental colonialism, that each country has the right and sovereignty to decide which living marine resources to use in a sustainable manner, and that the whaling issue is a human rights and food security issue.[12]

For proponents of sustainable use, the whaling issue is not a symbol of environmental protection but of the principle of sustainable use of living resources.

The Future of Japanese Whaling and the Future of the IWC

In July 2019, Japan resumed commercial whaling in its territorial waters and exclusive economic zones. There are two major issues to be addressed in the future of the commercial whaling. One is to demonstrate to the international community with a high degree of transparency that the whaling conducted after leaving the IWC is not lawless and does not have a negative impact on the whale populations, and it is based on scientific evidence and sustainable harvest management, which is being observed under appropriate monitoring and control measures. Another issue is to reestablish commercial whaling as a self-sustaining economic activity, which was resumed after a long interval of about 30 years.

For the former, catch limits for the three whale species of minke, Bryde's, and sei whales, which are confirmed to be abundant, were calculated using the RMP, a calculation method developed by the IWC itself that does not adversely affect the whale stocks. In addition, monitoring and enforcement measures include DNA analysis and registration of whale meat by the DNA information to prevent poaching and smuggling, monitoring of whaling ship locations using satellites, and boarding of supervisors on board whaling vessels.

Time will tell if commercial whaling will be economically viable in the future, but the cost of commercial whaling will be lower than that of research whaling (e.g., the research whaling program has to operate even in low-whale–density areas to obtain statistically unbiased data, while commercial whaling can operate mainly in high-density areas). Substantial economic advantages are also expected for commercial whaling because more flexibility is allowed in the sales of whale meat than the research whaling i.e., in the highly public (governmental) whale research program, fair distribution of whale meat was required, resulting in a wide and thin distribution, which caused problems in product assortment, quality assurance, and new product development.

Whaling activities and whale meat consumption are inherently local in nature. Although whale meat was consumed nationwide during the food shortages after World War II and when whale meat was used in school lunches, the major whale meat consumption areas in the past and present are Kyushu and Tohoku regions.

The highest per capita consumption of whale meat is in Nagasaki Prefecture, about four times the national average. Saga, Miyagi, and Yamaguchi prefectures follow. While it may be important to promote whale meat consumption in the heavily populated Tokyo area, given the limited total supply, it may be more desirable from both a business strategy and a local development perspective to concentrate sales promotion in areas where whale meat consumption has traditionally been high.

From an international perspective, what should Japan's whaling policy be in the future? As stated in the Chief Cabinet Secretary's statement on 26 December 2018, Japan will continue to participate in the IWC as an observer even after withdrawal and will participate in discussions on whaling issues in the international community. This appears to be a highly defensive response to Article 65 of the United Nations (UN) Convention on the Law of the Sea, which stipulates that the conservation, management, and study of cetaceans shall be conducted "through the appropriate international organizations,"[13] in order to reduce the risk of further litigation. However, rather than this defensive response alone, an active response is needed to build a new international principle and framework for the sustainable use of cetaceans as a living marine resource, because the IWC has effectively abandoned this mission. It is important for Japan to clearly demonstrate by its actions that it is not conducting commercial whaling on its own with its back turned to the international community. Japan has in fact proposed in the past the establishment of a second IWC and the establishment of a new regional international organization in the North Pacific. The author hopes that the withdrawal from the IWC will be an opportunity for these initiatives to be launched once again.

What will happen to the IWC after Japan's withdrawal? There are various possible scenarios, one of which is that Japan will be followed by a large number of other countries and the IWC will lose its reason for existence. A few years have already passed since Japan announced its withdrawal in late 2018 at the time of this writing, and no signs of this mass withdrawal have appeared. Of course, we cannot rule out the possibility of a gradual decline in the number of member countries or an increase in withdrawals around the time of the future IWC Annual Meetings. However, there are several reasons to speculate that a mass withdrawal will not happen. First, Japan is the only country that would be harmed by staying in the IWC under the current situation. Japan is bound by the IWC's moratorium on commercial whaling and has suffered the actual harm of not being able to resume commercial whaling, which is its national policy, but the other whaling nations, Norway and Iceland, have been legally conducting commercial whaling through the objection process (Norway) and in the case of Iceland the reservation process when they joined the IWC. Indigenous whaling nations Russia and Greenland have also been granted catch quotas legally under the IWC. Other countries that support sustainable use of whales have no desire to start commercial whaling, at least in the short term. In other words, there is no reason why any country other than Japan should actively pursue the path of withdrawal. Another reason is that, as noted above, the policy objectives of the sustainable use advocacy countries focus on upholding and promoting the principle of sustainable use in the international community, and this will not change with Japan's withdrawal. The withdrawal is more on the side of the many anti-whaling countries that have joined only to secure votes against whaling. After Japan's withdrawal, the significance of these anti-whaling countries in the IWC will be greatly reduced.

Another scenario is that the IWC will be transformed into an institution devoted to whale protection even more than it is now because of the withdrawal of Japan, a vocal supporter of sustainable use of whales. The IWC has already spent considerable time and resources on reducing and preventing the bycatch of whales by fisheries, assessing and preventing the negative impacts of climate change and other factors on whales, and animal welfare issues concerning whales, and it is not difficult to predict that this trend will further accelerate because Japan's withdrawal will lead to reduced or no discussions on scientific whaling and a catch quota request proposal for Japan's small-type coastal whaling program. Perhaps, realistically, the situation will be somewhere in the middle of these scenarios, that is, fewer IWC members will attend the IWC meetings (and/or pay their contributions to the IWC budget) and IWC will further become more of a whale protection agency. As the IWC's attention shifts to whale protection, the IWC's interest in and need for science for whale resource management will diminish.

However, the utilization of marine mammals, including whales, is taking place in a large number of countries, and in many cases marine mammals migrate widely beyond each country's 200 nautical mile waters, there is a clear need for their conservation and management. According to Robarts et al., marine mammals, including whales, were consumptively utilized in 114 countries of the world between 1990 and 2009.[14]

Given the IWC's de facto refusal to manage the utilization of whales, it is incumbent on Japan, which has withdrawn from the IWC, to propose and implement a concept for the international conservation and management of marine mammals.

Conclusion

The whaling issue is not a unique and isolated problem. Within the international community, there are many debates and conflicts over the relationship between animals and humans, and they are affecting our daily lives and food security. The whaling issue involves all elements of the animal-human relationship, including science, law, economics, culture, ethics, and emotions, and therefore there is a strong sense among international stakeholders involved in the issues of conservation and management of living resources that the whaling issue is a symbol and frontline of these issues. This is the viewpoint that the whaling issue should be addressed from a broader perspective, rather than from the perspective/interests of a few concerned parties, because if the whaling issue is handled incorrectly, other biological resource management issues will be adversely affected. This is the whaling as a bulwark theory. However, from the perspective of the "environmentalists," including animal rights proponents, the viewpoint is to promote the expansion of their influence through the whaling issue.

If an animal is in fact endangered, it is only natural to try to avoid extinction or to increase its population through a ban on harvest/catch or international trade.

However, the IWC even denies the use of whale species such as minke whales, which the Scientific Committee recognizes as abundant, with a catch quota calculated according to the IWC-approved scientific formula. Even the Secretariat of the Convention on International Trade in Endangered Species of Wild Fauna and Flora[15] acknowledges that there are some species listed in its Appendix I (endangered) that are not biologically endangered. There is no shortage of examples of human livelihoods being sacrificed to protect animals regarded as charismatic.

While it is important to protect pristine nature, the development and use of nature for food, energy, and other resources is also essential for human survival. Use and protection of nature is often viewed as binary and mutually exclusive and therefore has given rise to various conflicts. However, the principle of sustainable use, using nature while preserving nature, strikes a balance between the two and is a principle supported by many international organizations and nations, including the UN. The concept of charismatic megafauna and the idea of environmental protection are essentially two different things, but some argue that the use of charismatic animals is environmentally destructive. This is also symbolized by the whaling issue.

In the IWC, countries that support whaling call themselves supporters of sustainable use. This is in recognition of the fact that the whaling issue has much in common with the sustainable use of all living resources and the movement to restrict or deny it, and that the outcome of the whaling issue will affect the broader issue of sustainable use of living resources. Although Japan has withdrawn from the IWC and resumed commercial whaling, the issues surrounding the sustainable use of living resources will continue to grow in importance. Japan will continue to participate in the IWC as an observer in order to continue to address "another policy goal" that the whaling issue represents, and sustainable use advocates hope that Japan will continue to show leadership.

Notes

1 International Convention for the Regulation of Whaling 1946, Schedule.
2 IWC Circular RG/ VJH/19814, 26 May 1993.
3 Schedule, Art 10(e).
4 See for example Malgosia Fitzmaurice and Dai Tamada, *Whaling in the Antarctic. The significance and the implications of the ICJ judgment* (Brill 2016).
5 See for example The Sydney Morning Herald, 'Rudd foolish on whaling stance: Turnbull' *The Sydney Morning Herald* (Sydney, 21 May 2007) https://www.smh.com.au/national/rudd-foolish-on-whaling-stance-turnbull-20070521-dxm.html accessed 20 April 2023.
6 See for example Oliver Milman, 'WikiLeaks cables reveal Australian 'middle power diplomacy' on green issues' *The Guardian* (London, 8 September 2011) https://www.theguardian.com/environment/2011/sep/08/wikileaks-australia-environmental-relations accessed 20 April 2023.
7 Whaling in the Antarctic (Australia v. Japan: New Zealand intervening) [2014] Judgment, ICJ Reports [2014], 226. https://www.icj-cij.org/sites/default/files/case-related/148/148-20140331-JUD-01-00-EN.pdf accessed 20 April 2023.
8 See Sellheim and Morishita in this volume.
9 Ministry of Foreign Affairs of Japan, International Court of Justice. Whaling in the Antarctic, 31 March 2014. https://www.mofa.go.jp/ecm/fsh/page2e_000012.html accessed April 2023.

10 Yoshimasa Hayashi, Policy towards the Future Whale Research Programs Statement by Minister for Agriculture, Forestry and Fisheries, the Government of Japan, 18 April 2014. https://www.jfa.maff.go.jp/e/pdf/danwa.pdf accessed 20 April 2023.
11 On this very issue, see Richard W Bulliet, *Hunters, herders and hamburgers. The past and future of human-animal relationships* (CUP 2005).
12 See the intervention of the Commissioner of Antigua & Barbuda on the tabled resolution on food security at IWC68.
13 United Nations Convention on the Law of the Sea 1982.
14 Martin D. Robarts and Randall R. Reeves, 'The global extent and character of marine mammal consumption by humans: 1970–2009', (2011), *Biological Conservation* 144, 2770.
15 Convention on International Trade in Endangered Species of Wild Fauna and Flora 1973.

Part I
Institutional Implications

3 Exit Japan, Exit International Whaling Commission?

Steinar Andresen and David Aarvik Nese

Introduction

For centuries, whaling has been an important industry and a classic example of the tragedy of the commons. To rectify this development, the International Whaling Commission (IWC) was established. However, the IWC initially failed, and depletion continued. Over time, the IWC changed fundamentally into a protectionist body. Paradoxically, when whaling was a sizeable industry, the IWC was exceedingly weak. However today, hundreds of participants meet while whaling as an industry has vanished. This begs the question whether the IWC in its present form has any relevance. This question is becoming even more relevant in light of Japan's exit from the IWC because Japan has been the most import contracting government of the IWC regarding key dimensions of the organization's work.

In this chapter we trace the development of the IWC based on novel datasets that allow for statistical insights into the organization and the role Japan has played therein.[1]

Methodology

This chapter presents a novel dataset, which includes harvest data, delegation data, and financial contributions, providing a new perspective on IWC. The dataset was compiled from several sources, including IWC's summary database (v7.1),[2] IWC reports (both Commission and Scientific Committee (SC) reports), IWC's website, and IWC financial reports.

Delegation data was collected from IWC meeting reports, counting the number of delegates from each delegation, categorizing each observation by meeting year, location, member, observer-status (or not), organization (or not), and country (or not). The report year was also included for future reference. Observations span from 1949 to 2021, totalling 5,299 observations. Observers were not further categorized, as the IWC has used different categorizations over the years.

The same procedure was applied to collect SC data, counting the number of delegates from each delegation and coding whether the delegation was a member or a country. The report year was also included for future reference. Observations span from 1971 to 2021, totalling 1,511 observations. This timeframe was selected because 1971 is the first year in which the SC has its own reports in the IWC archive.

DOI: 10.4324/9781003250814-4

IWC member data was retrieved from the IWC website, identifying when a member adhered to the IWC. However, as some countries have joined, left, and re-joined the IWC, this information was cross-checked in IWC reports for those countries, creating a panel dataset for all members. Each row represents a 'member-year', resulting in 3,087 observations.

Financial data was compiled by creating a dataset for each country-year and noting their financial contributions. Data from 2009 to 2021 was used due to the availability of IWC archives containing financial reports, resulting in 1,236 observations.

All of the datasets were merged into a single dataset comprising 7,541 observations.[3] The subsequent sections of this study present the results of our analysis and demonstrate how this dataset sheds new light on the dynamics of the IWC and its history of successes and challenges.

Whaling and the Development of the IWC

Catch

Whaling has been conducted for more than a thousand years and was initially pursued exclusively in coastal areas.[4] Historically, all major sea-going nations have also been whaling nations. During most of the 19th century, the US was the major actor, but Norway and the United Kingdom took over as the leading whaling nations when modern industrial whaling started in the Southern Ocean, early in the 20th century. The large baleen whales were taken exclusively for their whale oil, and the pre-IWC history of whaling is a classic example of the tragedy of the commons.[5] There were international efforts to regulate whaling in the 1930s, but without any effect.[6]

Whaling statistics date back to 1900.[7] As shown in Figure 3.1, whaling was a miniscule industry at that time, with only a few thousand whales taken annually. When pelagic whaling moved into Antarctic waters, the number of whales caught increased rapidly, although there were strong fluctuations. By the mid-1930s, the annual catch had increased to close to 65,000 whales. The reason for the dramatic subsequent decrease in catch was due to World War 2 when most whaling ships were used in World War 2 rather than for catching whales. After the war, whaling again increased rapidly and at the time of the establishment of the IWC in 1946, some 50,000 whales were caught annually.[8]

The IWC was initially not able to halt the depletion of whales. As Figure 3.1 illustrates the level of catch increased dramatically and reached an 'all time high' of more than 80,000 whales caught in the early 1960s. Moreover, the vast majority of these immense hauls during and before the 1960s were largely directed at the largest and endangered baleen whales, e.g. fin, humpback, sperm, and blue. The major reason was that the short-term economic interests of the major whaling nations dominated, and there was a fierce competition between them to catch as many whales as soon as possible in what has been characterized as the 'whaling Olympics'.[9]

From the mid-1960s, however, the level of catch has decreased dramatically, and it was down to some 10,000 whales when the moratorium came into force in

Figure 3.1 Total harvest of all whales; the figure displays the annual harvest of all whale species, including blue, pygmy blue, fin, sperm, humpback, sei, Bryde, common minke, antarctic minke, gray, bowhead, right, and unspecified whales, from 1900 to 2021. Each dot on the chart represents a year, and the data is sourced from the IWC's DB 7.1. It is important to note that the figure does not differentiate between the size and species of the whales, providing a general overview of the development of whaling with a focus on the total number of whales harvested.

the 1985/86 whaling season. There are several reasons behind this dramatic decrease. One important reason was that the whalers were no longer able to catch the quotas due to depletion. Moreover, the scientific influence as well as management procedures improved significantly within a much institutionally strengthened IWC.

The majority of new contracting governments that joined since late 1970s did not regard this as enough, and a moratorium on commercial whaling in the form of zero catch quotas was adopted in 1982. The moratorium was almost fully implemented by 1986, and by the end of the 1980s it appeared that virtually all whaling would stop and whaling has been miniscule since its coming into force. Still, more than 50,000 whales have been caught during this period of some 35 years. As detailed in Figure 3.2, three types of whaling have been conducted since the moratorium came into force. During most of this period, the controversial special permit scientific whaling, conducted primarily by Japan, amounts to some 18,000 whales. Commercial catch amounts to more than 20,000 whales, conducted mostly by Norway. Aboriginal Subsistence Whaling, permissible by the IWC, is conducted primarily by the US, Russia, Bequia (Saint Vincent and the Grenadines), and Denmark (Greenland) and represents the smallest fraction of the three. The year with the highest catch was in 2005, but since then it has been gradually reduced.

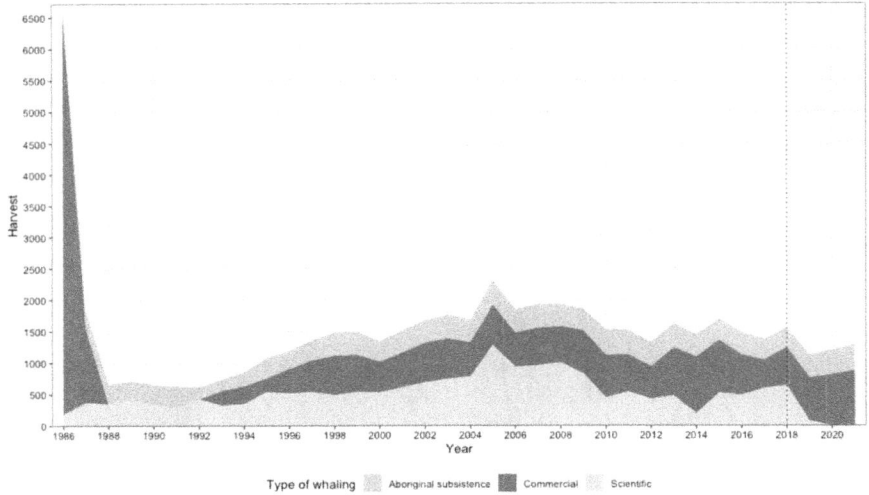

Figure 3.2 Types of whaling 1986–2021; the figure depicts the annual harvest of whales categorized by the type of whaling conducted after the implementation of the moratorium. Notably, in 1986, most of the harvest was attributed to unregistered whaling by the USSR. The data is sourced from the IWC's DB 7.1, using their own categories. It's worth mentioning that the 'Other' category, which includes illegal whaling activities and whaling by non-members, is excluded due to its relatively small size (only 388 whales from 1986 to 2020). Additionally, 2018 marks Japan's final year in the IWC, and in 2019, their scientific whaling was re-categorized as commercial.

There is hardly any reason to expect that the catch of whales will increase in the future. First, Iceland has hardly caught any whales the last few years and has been somewhat unclear whether harvest will continue or not beyond 2023.[10] As to Norway, catch has been stable at around some 500 animals annually, and there is no reason to expect that this will rise as demand for whale meat is very modest and there is a long time since the quota was fulfilled.[11] Also, scientific whaling has ceased after Japan's exit from the IWC, and its catch in the Southern Ocean has ceased and only very modest catch has been conducted within its economic zone.[12] As in Norway, very modest demand makes it quite unlikely that it will increase much in the future.[13]

Also important from the perspective of the future of the IWC is that the Norwegian commercial catch is not regulated by the IWC, and now the same goes for Japan's commercial catch after its exit. It is an open question whether Iceland is interested in staying as a member of the IWC if it will cease to be a whaling nation. With only the miniscule Aboriginal whaling under its regulatory purview, the number of delegates of various kinds in the Commission and the SC as will be specified in the next section is much higher than the number of whales caught for the purposes of Aboriginal Subsistence Whaling.

Participation

How was it possible that an organization that originally worked to protect the whaling industry shifted to a conservation organization? Figure 3.3 shows the number of members of the IWC and the ratio between whaling and non-whaling nations. Initially, the overwhelming majority of its members were whaling nations. This is in line with international management bodies for marine living resources where participation is usually restricted to stakeholders only with manifest material interests. In contrast, the IWC is open to all interested states by virtue of its open accession clause. While the number of whaling nation members for various reasons was reduced, Figure 3.3 shows that the number of non-whaling members has skyrocketed since the late 1970s. This increase was particularly steep in the years leading up to the adoption of the moratorium. Then it was rather stable with less than 50 member states for the next 20 years or so. Early in this century, there was a new sharp increase and while there were almost exclusively anti-whaling nations in the first steep increase, this time pro-whaling nations constituted most of the new members. In fact, in 2006, there was a 33–32 majority for the pro-whaling section. However, this trend has not continued. Membership continued to

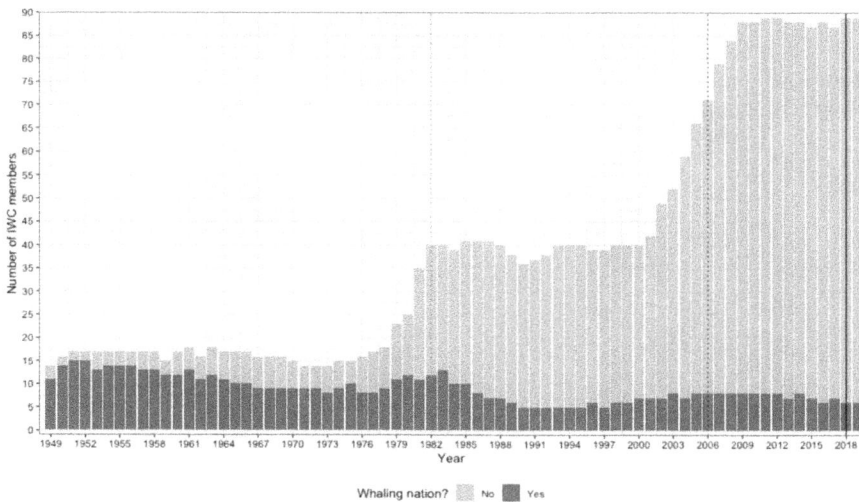

Figure 3.3 IWC whaling and non-whaling members; the figure displays the development of member states and whaling nations within the IWC from 1949 to 2019. The number of member states has remained stable after 2019. Each bar represents a given year, with the total number of member states shown in the bar. The dark colour in the bar represents the absolute share of member states that are whaling nations, while the light colour indicates the number of member states that do not engage in whaling. A light dotted line marks the year when the moratorium on commercial whaling was implemented, and a black dotted line denotes the St. Kitts and Nevis Declaration, which was a 33-32 vote to lift the moratorium.[14] Finally, a black line illustrates Japan's final year as a member of the IWC.

increase until around 2010 and since then it has stabilized with some 80+ members, most of them anti-whaling, and the pro-whaling section is nowhere near the needed ¾ majority needed to lift the moratorium and it is hard to see that this will ever happen. The likelihood is now even smaller in light of the exit of the most prominent pro-whaling member, Japan.

What explains this fundamental turn-about by the IWC from being a whaling club to a protectionist club? A number of factors underlie this development; the fact that the whale was adopted as a symbol for the green – and animal right groups; this sentiment spread to key Western states, not the least the US; considering that they had no material interests at stake, it was a convenient way to stand forth as environmentalists; and finally, the 'policeman role' of the US in the effective implementation of the moratorium.[15]

More specifically, regarding the dramatic rise in participation, active recruitment by key actors explains a lot. In securing enough votes to adopt the moratorium, it is well documented that both the US and Greenpeace were creative and effective in this regard.[16] Japan played a similar role in the second active recruitment period.[17] Most of these new members of the IWC only send one member to the Commission meetings. The modest level of priority paid to these meetings by many members is also reflected in the level of participation in the SC meetings. Recall here that according to the ICRW, all decisions are to be based on scientific advice, the reason why this key committee was established in 1951. Initially, membership was very modest, and in the 1950s only a handful scientist took part.[18] In Figure 3.4, an account of the participation in this Committee is presented from 1971 to 2022. There

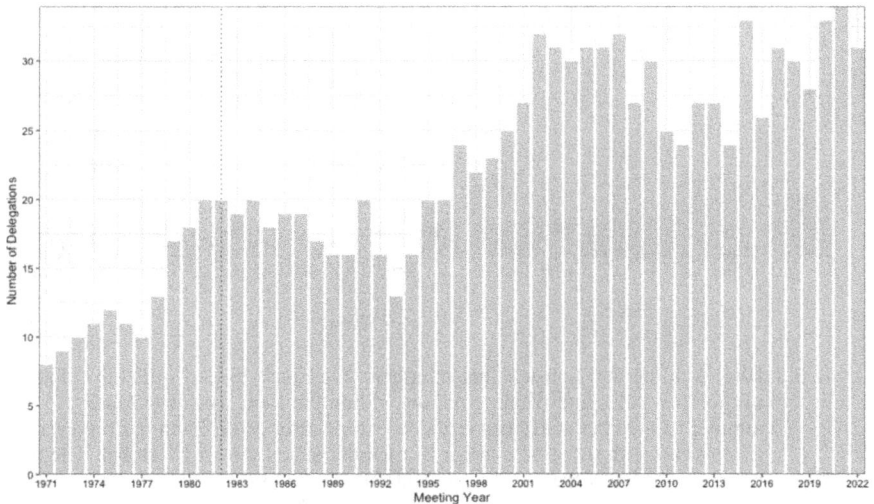

Figure 3.4 Member state delegations within the SC; the figure shows the development of member state delegations within the SC of the IWC. For the sake of readability, the figure excludes member organizations and observation organizations/nations. Additionally, the number of delegates from each delegation is not included in the figure.

has been a strong increase in participating member states here as well to the present level of slightly in excess of 30 delegations. Still, it indicates that more than 50% of the IWC members do not send any delegates to the SC at all. As will be detailed below, this is in stark contrast to the high number of participants sent by the key members of the Committee.

In this context, it is important to point to the fact that more recently the overwhelming majority of members in the SC are independent scientists, not representing any member states. Since the early 1990s, the Committee has suggested very advanced and cautious management procedures. If these procedures had been accepted by the member states, commercial whaling could have been resumed at a modest level.[19] However, the majority of IWC members have chosen not to listen to the scientists, indicating that decisions are based more on values and politics rather than science.[20]

Finally, let us take a brief look at the presence of non-state observing participants. During its first 20 years of existence, there were only a handful of them present, and the whaling industry was the most important actor.[21] As detailed in Figure 3.5, this changed dramatically from the mid- to late-1970s where their participation increased even more strongly than the rise in membership. These organizations were national, regional, or international environmental NGOs with significant budgets and human resources with a stated goal to end commercial whaling.[22] For a long period, they were far more plentiful than member states present and more than

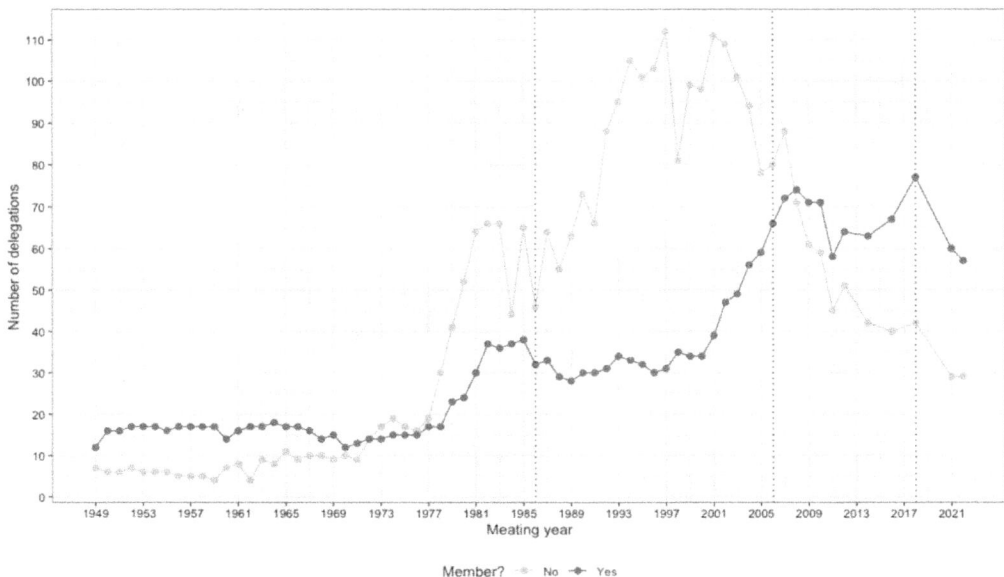

Figure 3.5 IWC member state delegations; the figure displays the number of delegations from member states (represented by the dark bars) and observers (represented by the light bars). Each dot is a meeting in IWC. The data used in the figure is sourced from the reports of the IWC.

a hundred of them have at times been present during the peak period around the turn of the millennium. As illustrated in Figure 3.5, the number of member states delegation was around 30–50 with around 2–3 delegates each, the environmental non-governmental organizations (ENGOs) could vary between 70 and 110 delegations with around 2 delegates each making IWC significantly lopsided against whaling. With these organizations, the US and other anti-whaling nations an unbeatable alliance was created by controlling the meetings, lobbying and pushing the neutrals to join the anti-whaling side. Hence, the pro-whaling nations were both outnumbered and outplayed.

Interestingly, however, during the last 20 years, their presence has been dramatically reduced and is now only about a third compared to the peak years. In our opinion, the explanation is quite straight-forward. The whole whaling issue is no longer prominent on the international political agenda as used to be the case in the 1980s and 1990s. It used to be a favourite in the Western media, showing dramatic pictures of the 'slaughtering' of whales. More recently, the IWC meetings go largely unnoticed by the international media. Even in Norway, where the issue was very high on the agenda, it has disappeared completely.[23] The meetings have turned into routine with little or no practical importance, and all efforts to break the stalemate and reform the IWC have failed, and Japan's attempts to get some positive response to its demands have failed similarly. As detailed in many of the other chapters in this book, this is why Japan finally decided to leave the organization.

Japan – the Most Important Member of the IWC?

We will not reflect further on Japan's exit in this chapter. Rather, we will – through a number of graphs and tables – illustrate the significance of Japan as a whaling nation and particularly its key role in the IWC. On this basis, we will reflect on the significance and effect of Japan's exit for whaling and particularly for the future of the IWC.

Japan's Whaling

Figure 3.6 shows Japan's development as a whaling nation since World War 2. It started at a very low level with a few thousand whales caught per year. The level of catch increased very rapidly during the next two decades and peaked in the mid-1960s with close to 30,000 whales caught. The figure shows, first, that Japan's catch peaks at the same time as the overall global catch. Second, that its catch was more than 30% of total catch at that time. Japan was also, together with the Soviet Union, the last dominating Antarctic whaling nation.

From the mid-1960s, however, as with total catch, the level was reduced rapidly. In fact, the level of catch at the time of the coming into force of the moratorium was down to the same level as the catch in the 1940s. Still, as illustrated in Figure 3.7, there was a dramatic fall in Japanese catch following the implementation of the moratorium, from almost 3,000 whales to only a couple of hundreds per annum. As Japan withdrew the objection to the moratorium due to pressure from the US, it could not do commercial whaling and still comply with the ICRW.[24]

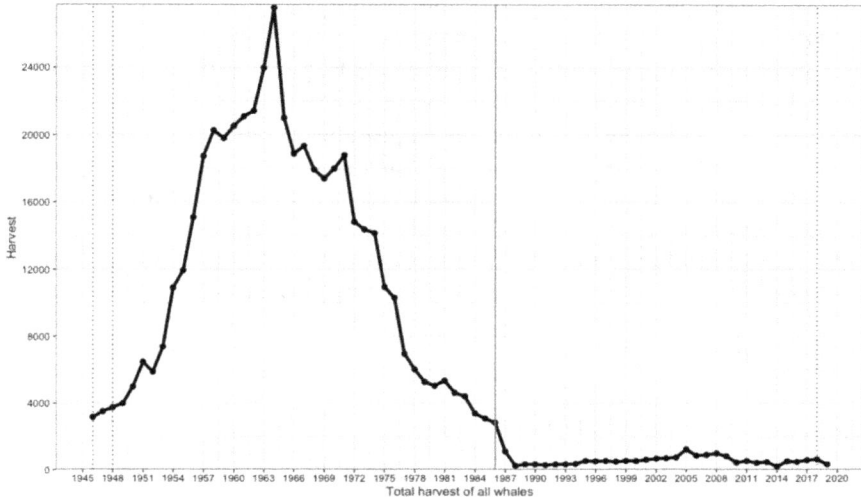

Figure 3.6 Japan's annual whale harvest; the figure depicts Japan's annual harvest of whales, including blue, pygmy blue, fin, sperm, humpback, sei, Bryde, common minke, antarctic minke, gray, bowhead, right, and unspecified species, aggregated from 1945 to 2019 from IWC's DB 7.1. It's worth noting that the figure does not differentiate between the sizes and species of the harvested whales, providing a general overview of Japan's overall whaling activities during this time period.

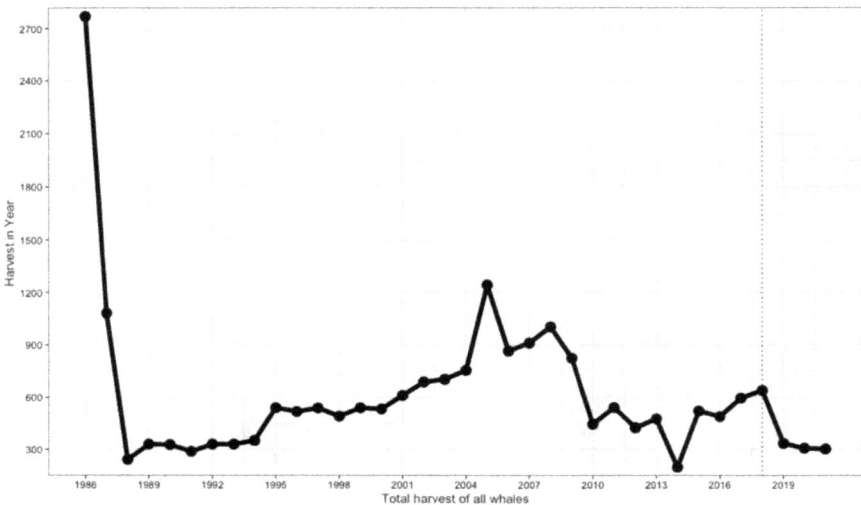

Figure 3.7 Japan's harvest of sei, Bryde, minke, and sperm whales 1986–2021.

Instead, it started the permissible scientific whaling. Japan holds that they simply hunted for scientific purposes to show that commercial whaling was possible. Their opponents have claimed that Japan is utilizing a loophole in the whaling convention.[25] As Figure 3.7 illustrates, the level of catch increased quite strongly from this very low level and some 1,200 were taken just after the turn of the century. This was about twice the level of commercial catch that Norway has ever taken after the moratorium. Although resolutions were regularly adopted against its whaling, this for long did not seem to affect Japan's level of catch. However, since the peak, Japan's catch has been significantly reduced, and the verdict of the International Court of Justice against one Antarctic whaling program no doubt contributed to this development.[26] During the last few years of Japan's IWC membership, only a few hundred whales were taken annually.

Japan – the Most Powerful Member of the IWC?

The information presented in the following graphs illustrates the dominant position of Japan in the IWC along a number of key dimensions. Moreover, it is also demonstrated that there are only a handful of nations truly active in the IWC, although there are 88 members.

Figure 3.8 gives an overview of the mean delegation size of IWC members in the Commission and the SC. The figure illustrates that Japan and the US have been

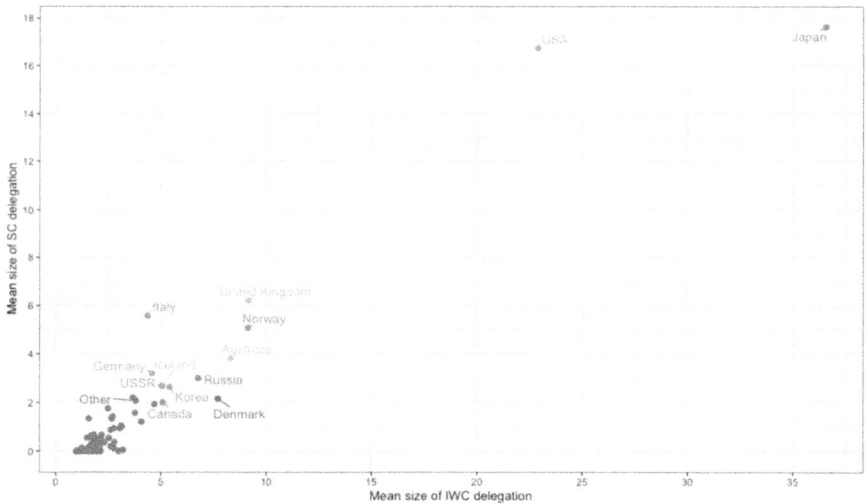

Figure 3.8 Mean delegation sizes within the IWC; the scatterplot depicts the mean delegation sizes of the IWC on the x-axis and the SC on the y-axis from 1971 to 2021. The figure highlights the delegation sizes of major players in the IWC and SC, including Japan, the USA, the UK, Norway, Italy, Australia, Denmark, and Germany. Additionally, each dot in the figure represents a country, with the last 86 countries marked in red to account for changes in IWC membership over time. The figure provides a clear visualization of the relative delegation sizes of key players within the IWC and SC.

'Participatory giants' in the IWC between 1971 and 2021.[27] It shows that Japan has an average of more than 35 participants to the Commission meetings and almost 18 participants in the SC. The US is quite far behind Japan in the Commission but quite close in the SC. Trailing very far behind with less than ten member delegations in the two forums are the United Kingdom, Norway, Italy, Australia, Denmark, and Germany. As the cluster shows in the left bottom section of the figure, the vast majority has very limited participation.

Figure 3.9 gives an overview of a combination of the mean harvest and contribution to the IWC budget in the period 2011 and 2019. Thus, this figure illustrates both vested material interests (catch) and the members' significance for the IWC as an effective organization. Here, Japan has the second largest economic contribution and the highest level of catch, with Norway coming very close along the latter dimension. Despite their Aboriginal catch being very modest, the US is the biggest contributor to IWC due to their considerable voluntary financial support. These voluntary contributions are largely earmarked to Aboriginal Subsistence Whaling and the IWCs whale watching project.[28]

Moreover, it is also worth noting the difference between the most prominent anti-whaling and whaling nations where the most prominent anti-whaling nations (i.e., Australia, the United Kingdom, Italy, and France) contribute significantly more than the whaling ones (i.e., Norway, Iceland, Denmark, and Russia).[29]

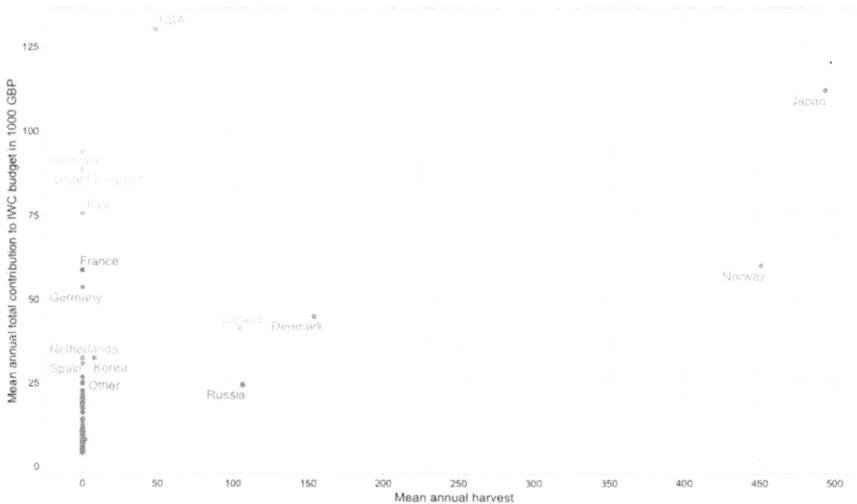

Figure 3.9 Mean annual harvest and contributions to IWC budget; the scatterplot shows the mean annual harvest (x-axis) and mean annual contribution to the IWC budget (y-axis) between 2009 and 2022. The figure highlights the discrepancy between whaling and anti-whaling nations, with the data revealing that whaling nations tend to have higher annual harvests and lower contributions to the IWC budget than anti-whaling ones. The figure uses voluntary contribution data from 2009 onwards, including both set and voluntary contributions, with total contributions adjusted for inflation using 2009 GBP as a baseline. The mean for Japan is based on data from 2009 to 2018. Contributions are adjusted by using "PriceR" that utilizes inflation rate data from the World Bank API.

As such, Japan sticks out as they harvest the most whales while also contributing way more than the other whaling nations to an international organization where they routinely have been shamed.

Another way of illustrating the significance of various actors for the IWC as well as showing their priority given to the IWC is to combine the number of delegates to the IWC and financial contribution. Figure 3.10 illustrates that Japan is the 'super-category' with the US in this regard as well. Other members worth mentioning, but trailing far behind, are Australia, the United Kingdom, Norway, Italy, Denmark, Iceland, and South Korea.

In short, these figures (three scatterplots above) clearly illustrate the key role that Japan has played in the IWC. The only other actor that matches Japan is the US with a handful of other fairly important actors trailing far behind. This group is composed of very vocal anti-whaling nations like the United Kingdom and Australia as well as pro-whaling nations like Norway and Iceland. To simplify somewhat, there are two giants, a handful of quite significant actors, while the vast majority is usually no more than followers and spectators.

Based on standard international relations and political science theory, considering the dominant role of Japan for the IWC, one should expect Japan to have considerable influence on decisions taken in the IWC. This can be illustrated through the famous notion that, 'votes count, but resources decide', as expressed by the Norwegian political scientist, Stein Rokkan.[30] This standard assumption seems

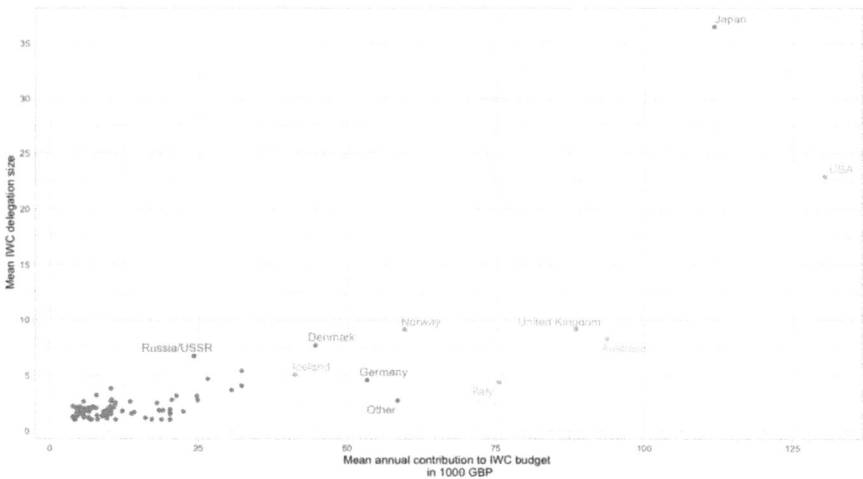

Figure 3.10 Mean delegation size and annual total contribution; this scatterplot illustrates the mean annual total contribution (x-axis) and mean IWC-delegation size (y-axis) between 2009 and 2021. The figure highlights the relationship between IWC delegation size and total contribution, with larger delegations tending to correspond to higher total contributions. The figure uses voluntary contribution data from 2009 onwards, including both set and voluntary contributions, with total contributions adjusted for inflation using 2009 GBP as a baseline. The mean for Japan is based on data from 2009 to 2018. Contributions are adjusted by using "PriceR" that utilizes inflation rate data from the World Bank API.

invalid for this rather peculiar international organization. Rather, decisive power is held by the majority of IWC members, with moderate interest and priority, through their votes cast when various anti-whaling decisions are taken. Thus, in this case, the quote can be turned on its head, 'resources count, but votes decide'. We argue this is the case because of the virtual absence of material interests present for the vast number of participants. Thus, participants working against whaling have been driven by norms and values rather than material interest. Hence, IWC can be interpreted as an exception in international politics where symbols, norms, and values are what matters, not material interest as in 'normal' international bodies for marine resource management.

Exit Japan – Exit the IWC?

We have shown that the exit of Japan will turn the IWC into a fundamentally new international organization. First, in terms of participation in Commission meetings, Japan's exit in numerical terms represents the same as if some 45 member states with 1-person delegation should leave the organization.[31] Also, obviously, the scientific input will likely be significantly reduced with Japan's exit. Despite that Japan still participate as an observer in both the IWC and the SC, it is not unlikely that Japan would rather shift focus to other forums (like the North Atlantic Marine Mammal Commission, NAMMCO) and rather use its vast resources there.[32]

Secondly, the atmosphere of this body may also be affected by Japan's exit. Japan has been the most important antagonist against the anti-whaling majority. Surely, controversies will still exit, but the conservation side will be even further strengthened (Figure 3.11).

A third effect, based on Japan's high economic contribution to the IWC, seems to represent a considerable reduction in the IWC budget with negative effects for its work in the future. As exemplified in the most recent SC meeting and subsequent report, where budgetary cuts were actively discussed to solve IWC financial deficit in the next budget.[33] However, income from membership contributions is not the only source of income for the IWC. As shown in the figure above, the last 10 years there has been a trend where anti-whaling states, especially the US and Australia, have contributed with voluntary earmarked contributions.[34] Potentially, this trend might continue compensating for the lapse of contribution from Japan in the future. Nevertheless, Japan's exit has forced the IWC to fill the financial gap while also handling the downward trend in member contribution. With Japan's exit, it is also possible that others would reconsider their membership or potentially downscale their contribution even more, as the motive to change Japan's position is now completely futile in IWC.

In all likelihood, the IWC will continue its conservation work in the foreseeable future as international organizations tend to live on, although their original purpose has ceased to exist, also coined 'dead letter regimes'.[35] Our prediction is, however, that the attention paid to its working and its future relevance will continue to decrease as the issue is solidly placed on the backburner of attention for policymakers, media, as well as the green movement. Today there are a lot of serious

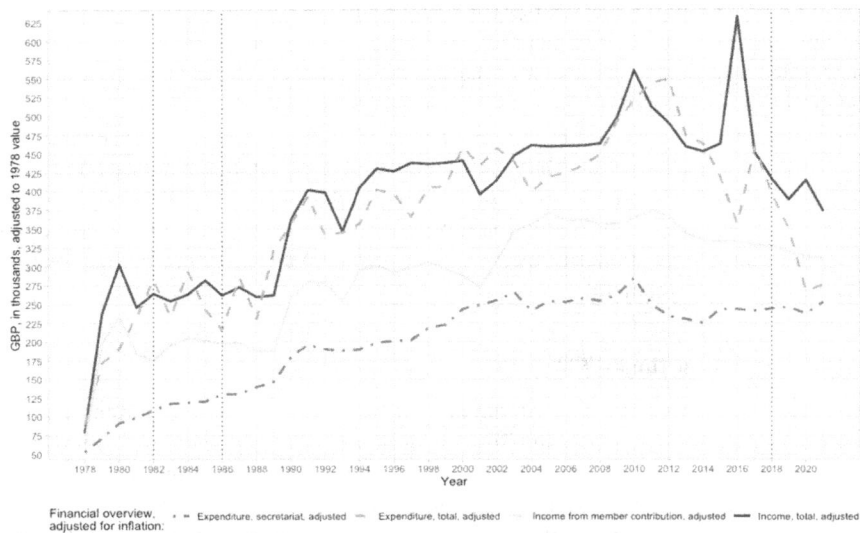

Figure 3.11 Financial overview of the IWC; this figure presents the annual financial overview of the IWC between 1978 and 2021, adjusted for inflation using 1978
GBP as the baseline. The dashed lined indicates the total expenditure and the
dot-dashed line the expenditure for the secretariat, while the dark line represents the income, which includes both the total income and the income from
member contributions (voluntary contributions are excluded) in the light line.
The data used in this figure is sourced from the IWC's financial statements.
Contributions are adjusted by using "PriceR" in R that utilizes inflation rate
data from the World Bank API.

threats to whales coming from a number of sources, but commercial whaling is not
among these threats.[36] Thus, ideally the IWC should change its charter to deal with
these threats. Alternatively, it could be transformed into a small Aboriginal whaling
body. However, based on all failed efforts to reform the IWC, we do not believe
that any of these scenarios will unfold.

Notes

1 Based on two novel datasets with more than 7,400 observations and more than 50 vari-
   ables, we track the development in IWC in terms of whale harvest, delegation sizes, and
   financial contribution. For more on the dataset, see appendix.
2 C. Allison, "IWC summary catch database Version 7.1." available from the International
   Whaling Commission (Cambridge 2020).
3 The dataset is available upon request from the authors.
4 Nils Chr. Stenseth, Alf Håkon Hoel and Ingunn B. Lid, *Vågehvalen: valgets kval*, (Gyldendal 1993).
5 Garret Harding, 'The tragedy of the commons', (1968), *Science* 162(3859), 1243.
6 Steinar Andresen, "The effectiveness of the International Whaling Commission",
   (1993), *Arctic* 46(2), 108.
7 C. Allison, "IWC summary catch database Version 7.1", available from the International
   Whaling Commission (Cambridge 2020).

8 Patricia Birnie (ed.), *International regulation of whaling: From conservation of whaling to conservation of whales and regulation of whale-watching* (Oceana 1985); John N Tønnessen and Arne O. Johnsen, *The history of modern whaling* (UCP 1982).

9 Alf Håkon Hoel, 'The International Whaling Commission 1972–1984: New Members, New Concerns' (Oslo, Fridtjof Nansen Institute 1985).

10 Iceland's Minister of Fisheries stated in 2022: "There are few justifications to authorise the whale hunt beyond 2024". See AFP, "Japan's return to commercial whaling in 2019 has left few buyers for Iceland's meat", *The Guardian*, (London: 4 February 2023). Conversely, Iceland is still communicating through NAMMCO that they are maintaining their pro-whaling policy. See NAMMCO, "Opening statement Iceland", https://nammco.no/wp-content/uploads/2022/09/nammco-opening-statement-iceland-13092022-js.pdf accessed 21 May 2023.

11 The quotas have been between 600 and 1,300 the last 20 years but have never been reached. See Regjeringen, "Lavere vågehvalkvote i 2022", https://www.regjeringen.no/no/aktuelt/lavere-vagehvalkvote-i-2022/id2901488/ accessed 21 May 2022.

12 The Japanese pelagic harvest is around double the size of the coastal harvest.

13 David Aarvik Nese, *Ambiguity, free-riding, or principles? Explaining Norwegian commercial whaling* (UIO 2022).

14 As a result of Japan's active recruitment, IWC had a voting to lift the moratorium in 2006. This resulted in the *St. Kitts and Nevis Declaration* which confirmed IWC initial purpose as organization to promote whaling interests. The resolution can be found here: IWC, "2006-1, St. Kitts and Nevis Declaration", https://archive.iwc.int/pages/download.php?ref=2081&ext=pdf&alternative=2993&noattach=true accessed 18 April 2023. For more on this, see Mike Ilif, "Modernization of the International Convention for the regulation of whaling", (2008), *Marine Policy* 32, 402.

15 Arne Kalland, *Unveiling the whale. Discourses on whales and whaling* (Berghahn 2011); Jennifer Leigh Bailey, 'Arrested development: The fight to end commercial whaling as a case of failed norm change', (2008), *EJIR* 14(2); Jennifer Leigh Bailey, 'Norway, the United States, and commercial whaling', (2009), *JED* 18(1).

16 Nicole Deitelhoff and Lisbeth Zimmermann, 'Things we lost in the fire: How different types of contestation affect the robustness of international norms', (2020), *ISR* 2(1), 51.

17 Jonathan R Strand and John P Tuman, 'Foreign aid and voting behavior in an international organization: The case of Japan and the International Whaling Commission', (2012), *Foriegn Policy Analysis* 8(4), 409.

18 Tore Schweder, "Regulering av hvalfangst – vitenskap og politikk", in Nils Chr. Stenseth, Alf Håkon Hoel and Ingunn B. Lid (eds.), *Vågehvalen: valgets kval*, (Gyldendal 1993), 291–310.

19 IWC, "The revised management procedure – further information", https://iwc.int/management-and-conservation/rmp/rmp2 accessed 18 April 2023.

20 Steinar Andresen, "Science and policy in the International Whaling Commission", in M.J. Paterson (ed.), *Contesting global environmental knowledge, norms, and governance* (Routledge 2019), 110; IWC, "2006-1, St. Kitts and Nevis Declaration", https://archive.iwc.int/pages/download.php?ref=2081&ext=pdf&alternative=2993&noattach=true accessed 20 April 2023.

21 For instance, the "International Association of Whaling companies" was an observer over several years during the 1950s.

22 For instance: Greenpeace, Sierra Club, American Cetacean Society, Fundacao Brasileira para a Conservacao da Natureza. If interested in the complete list of over 150 organizations, reach out to access the dataset, or see IWC meeting reports.

23 David Aarvik Nese, *Ambiguity, free-riding, or principles? Explaining Norwegian commercial whaling* (UIO 2022).

24 Jennifer Leigh Bailey, 'Arrested development: The fight to end commercial whaling as a case of failed norm change', (2008), EJIR 14(2), 319; S.D. Murphy, "U.S. Sanctions against Japan for Whaling", The American Journal of International Law 95(1), 149.

25 Michal Kolmas, "International pressure and Japanese withdrawal from the International Whaling Commission: when shaming fails", (2020), *Australian Journal of International Affairs* 75(2), 197.

26 The verdict is against "JARPA II". See Whaling in the Antarctic (Australia v. Japan: New Zealand intervening) Summary of the Judgment of 31 March 2014 (International Court of Justice).

27 The mean for Japan is estimated by using data from 1971 to 2018.

28 See IWC financial statements for more on what the voluntary contributions go to.

29 The vast majority has so marginal significance that they are not worthy of mentioning.

30 Stein Rokkan, 'Norway: Numerical democracy and corporate pluralism', in Robert Dahl (ed.), *Political oppositions in Western Democracies* (YUP 1966), 105.

31 After 2000, i.e., the last 15 meetings, Japan's mean delegation size has been 46 delegates, with the range between 29 delegates in 2009 and 73 in 2005.

32 NAMMCO, "Opening Statement Japan", https://nammco.no/wp-content/uploads/2023/03/jpn-opening-statementnammco30-2023_set_rev_240323-1.pdf accessed 22 April 2023.

33 Scientific Committee, "Report of the Scientific Committee (SC68D)", 187, https://archive.iwc.int/pages/download.php?direct=1&noattach=true&ref=19447&ext=pdf&k= accessed 18 April 2023.

34 A research consortium aimed to "maximise conservation-orientated outcomes for Southern Ocean cetaceans", (IWC, "The Southern Ocean Research Partnership (IWC-SORP)", https://iwc.int/scientific-research/sorp accessed 18 April 2023.

35 Marc A Levy, Oran R Young and Michael Zürn, 'The study of international regimes' [1995] EJIR 1(3).

36 IWC, "Introduction to Population Status", https://iwc.int/about-whales/population-status accessed 18. April 2023; IWC, "On World Wildlife Day, scientists highlight the threat of cetacean extinctions – and importance of acting now", https://iwc.int/resources/media-resources/news/on-world-wildlife-day-scientists-highlight-the accessed 18 April 2023.

4 As One Door Closes, Does Another Open? Assessing the Future of the Protectionist Agenda at the International Whaling Commission Post-Japan's Withdrawal

Cameron Jefferies and Heather Stock

Introduction

Whaling activity in Japan, as with many other island and coastal States, can be traced back several centuries.[1] The Japanese commercial whaling experience, which features mechanized pelagic whaling techniques, has been tracked to the Russo-Japanese War (1904–1905), where the Japanese navy captured four Russian whaling ships.[2] These Norwegian-built Russian ships were equipped with the latest industrial whaling technology and provided the foundation for a Japanese whaling fleet and network of whaling stations that hunted migrating whales in Japan's coastal waters. Japan's whaling industry rapidly developed and on May 1, 1909, six Japanese whaling companies merged to form the Tōyō Hogei whaling company, which controlled the majority of Japanese whaling ships.[3] This merger "stabilised the whaling industry for some time and allowed the [Japanese] Fisheries Agency to set a maximum limit of 28 whaling boats" in an effort to avoid overexploitation.[4] Japan's whaling experience from the first half of the 20th century is not unique; indeed, it is reflective of the general—and rapid—onset of industrial-scale pelagic whaling that triggered the corresponding need for a coordinated international legal response.

The ICRW[5] is one of the oldest international wildlife-related conventions. The ICRW established and empowered the IWC to operate as the primary forum for managing whale conservation. This regime's ability to steward conservation of the world's great whale species has, however, been limited by the ICRW's language and the IWC's contested operations. The ICRW's purpose is "to provide for the proper conservation of whale stocks and thus make possible the orderly development of the whaling industry"[6] yet it has rarely achieved either one of its dual objectives: the "proper conservation" of whales or the "orderly development" of whaling. To the modern observer, the ICRW/IWC's history can be divided into two management phases: Phase I (1946–1982), being the period in which the IWC (unsuccessfully) attempted to use market-based measures to avoid collapse of the whaling industry from overexploitation; and Phase II (1982–present), being the

DOI: 10.4324/9781003250814-5

period in which a *de facto* permanent ban on commercial whaling has prohibited most forms of whaling and a shift in IWC operations towards anti-whaling preservationist measures occurred.[7]

For the past three decades, conflict between pro-whaling and anti-whaling Contracting Governments to the ICRW/IWC has prevented the effective international cetacean conservation. Historically, these tensions manifest, as ineffective quota-setting,[8] shortcomings in monitoring and enforcement,[9] continued prosecution of "Special Permit" whaling, and the repeated denial of new whale sanctuary designations.[10] The pressures that have fractured the ICRW/IWC's functionality have been at the fore since 1982 when the IWC instituted a 10-year commercial whaling moratorium, which prompted a number of withdrawals and protests from pro-whaling countries.[11]

Despite the moratorium, in 1985, Japan's Minister of Agriculture, Forestry, and Fisheries, Moriyoshi Sato, stated that his "Government would do its utmost to find out ways to maintain the nation's whaling".[12] Starting in 1987, Japan controversially began to utilize Article VIII "Special Permit" whaling to support its national Japanese Whale Research Program Under Special Permit ("JARPA I") and the prosecution of pelagic whaling. This program conflicted—at least superficially—with the IWC's shifting focus to whale protections, as championed by influential states such as the United Kingdom and the United States of America, as well as various environmental non-governmental organizations (NGOs). The fractures between pro-utilization and pro-preservation Contracting Governments deepened and spread along political and cultural lines in the years following the temporary commercial whaling moratorium. The fact that the ban survived beyond the "comprehensive assessment" that was required to be completed no later than 1990 and has subsequently transformed into a *de facto* permanent moratorium has heightened this divide.[13]

Tension regarding Japan's use of Article VIII "Special Permit" whaling came to a head with the proceedings Australia initiated against Japan at the ICJ. The ICJ's decision in *Whaling in the Antarctic, Australia and New Zealand (intervening) v Japan*, concluded that Japan's whaling in the Southern Ocean under the second iteration of its Southern Ocean whaling program (Japanese Whale Research Program Under Special Permit, "JARPA II") violated the ICRW's moratorium on commercial whaling.[14] The Court specifically found that Japan's whaling was not conducted for the purposes of scientific research, as required by Article VIII. Post-*Whaling in the Antarctic*, Japan first revised its special permit whaling program in the Southern Ocean under a different name (New Scientific Whale Research Program in the Antarctic Ocean, "NEWREP-A") and subsequently introduced a proposed Schedule amendment before the IWC at its 2018 biennial meeting to end the moratorium and reinstate commercial whaling; this proposal was defeated by a vote of 41–27.[15] Following this defeat, Japan ultimately initiated its withdrawal from the ICRW/IWC regime in 2018, effective June 30, 2019.

Rand has astutely observed that the enduring contest between pro-whaling and anti-whaling states "[s]ave for the few instances where the IWC's membership has shifted to momentarily favour one side's arguments over another" has created a "perpetual impasse", impeding progress for either side.[16] The main consequence of continued in-fighting at the IWC, in our opinion, is that it hindered the international community's ability to focus its capacity on the most pressing anthropogenic threats to cetaceans: climate change, by-catch and ship strikes, marine pollution, entanglement, and habitat degradation.[17] The central proposal in *Marine Mammal Conservation and the Law of the Sea* was that, even amidst the conflict between the pro-whaling and anti-whaling factions, there might be room for a meaningful compromise that allowed limited commercial whaling in return for targeted action addressing these modern threats.[18] This outcome, in our opinion, still represents the best-case scenario. Unfortunately, such a compromise remains elusive and unlikely. In view of Japan's withdrawal from the IWC, we re-visit this argument and explore whether Japan's exit from the IWC closes the door on a truly global whaling compromise but opens another door that facilitates the transition of the ICRW/IWC to a third phase of its existence—one dominated by preservationist objectives.

Exploring the impact of Japan's withdrawal on the protectionist movement at ICRW/IWC requires us first to reflect on the ICRW/IWC's historical context (with a particular focus on Japanese whaling interests) and its management progression. Accordingly, this chapter proceeds in two parts. First, we introduce and survey the IWC's relevant history, beginning with its establishment to Japan's withdrawal. In so doing, we highlight the IWC's evolution through the first two phases of its existence. Second, we consider the probable effects of Japan's withdrawal from the IWC, with a focussed consideration of whether it, when placed in context of the IWC's increasingly progressive agenda, will enable the IWC's transition into a new management era, or Phase III, that is characterized as protectionist and focussed on combating the most pressing threats to cetacean survival by embracing holistic appreciation of the role of cetaceans within the marine environment. Our assessment concludes that Japan's withdrawal cannot fundamentally change the object and purpose of the ICRW or alter the three-quarters majority vote required to amend the regulatory Schedule; nor does it alter the IWC's limited regulatory capacity owing to the limitations of the ICRW's Schedule. While Japan's withdrawal negates the formal influence of one of the most vocal proponents of commercial whaling, expectations for some seismic shift towards a strong protectionist agenda must be tempered. Japan's withdrawal from the IWC has not opened the door to a strong global protectionist agenda; at best, it has nudged this door slightly more ajar.

The Relevant History of the *ICRW*/IWC

International Legal Efforts Pre-ICRW/IWC

The world's great whale species have expansive ranges and migratory routes and have thus proved particularly difficult to manage and conserve. The "depletion of one species of great whale after another is perhaps the most infamous example of human mismanagement of the Earth's natural resources"[19] and humankind's long history of utilizing whales for sustenance and as a resource source has fuelled the rise and collapse of various commercial enterprises—and, with them, whale populations.

The law of the sea is one of the oldest branches of public international law, but legal management of the world's great whales is a markedly 20th-century development that has failed to get in front of anthropogenic extinction pressures since its inception. The significance of the absence of international regulatory whale protection measures came to the fore in the 1920s when whaling nations, including the United States of America, Great Britain, Norway, and Australia, began to hunt for thousands of whales in the Antarctic region which, due to over-exploitation, almost resulted in the collapse of the global whaling industry.[20] Thus, the first international attempt to regulate whaling during the 1930s was approached with particular concern to Antarctic whaling. The 1931 League of Nations whaling meeting was the international community's first positive step towards international regulation, and the resulting *Convention for the Regulation of Whaling* was signed in Geneva in 1931 and came into force on January 16, 1935.[21] The convention applied to all waters, including the high seas and territorial waters, but only governed the hunting of baleen whales. The convention had little effect on Antarctic whaling because not all countries whaling in Antarctica participated. For example, both Germany and Japan were notably absent. However, the

> convention did establish the principle of international regulation of a common property resource in the high seas [...and] gave a legal framework for the voluntary production agreements entered into by the whaling companies after the gross over-production of [whale] oil in the 1930–31 Antarctic season.[22]

Recognizing the shortcomings of the 1931 convention, nine states, including Norway, the United Kingdom, and Germany, concluded a new treaty called the *International Agreement for the Regulation of Whaling* in 1937.[23] This agreement embarked on a more substantial regulatory agenda that forbade the taking of depleted right and gray whales, set minimum size limits for blue, fin, humpback, and sperm whales, and prohibited the capture of female whales accompanied by calves. Furthermore, the Antarctic pelagic whaling season was limited to December 8 – March 7, and large parts of the world's oceans north of 40°S latitude were closed to factory ship operations.[24] Japan, however, remained on the sidelines and did not adhere to the 1931 or 1937 Whaling Conventions. The official position for not acceding to the 1931 Whaling Convention was that the Soviet Union had not ratified the agreement and that Japan opposed agreement's prohibition on whaling for right whales in the North Pacific.[25] Regarding the 1937 Convention, despite concessions allowing whaling in the North Pacific that were designed to encourage Japanese participation, Japan did not accede.[26] Sumi notes that "[i]n order to remove international distrust, Japan expressed the intention to accede to them one year later, but the outbreak of the Second World War prevented it from doing so".[27]

In 1938, the Whaling Committee of a regional scientific advisory organization called the International Council for the Exploration of the Sea recommended a limit on the total allowable amount of oil produced by whaling. However, consensus on this limitation was not feasible because of the size and number of competing whaling fleets in the Antarctic. World War II (WWII) would change this, however, because "many [...] whaling factory ships had been sunk during the war and it was possible to set a maximum catch limit well below the pre-war catch level without placing unacceptable restrictions on the remaining fleets".[28] Thus, in 1944, the last significant pre-ICRW/IWC international whaling development occurred when an overall catch limit in the Antarctic was established.

The Inception of the ICRW/IWC and Phase I of Its Existence (1946–1982)

Despite the international accords of the 1930s, whale stocks continued to decline and, in 1946, the ICRW was concluded by 15 whaling states in Washington D.C.[29] The ICRW established the IWC, which met for the first time in negotiation coincided with the international community's realization of modern mechanized whaling's potential to dangerously over-exploit whale stocks. It purported to "ensure proper and effective conservation and development of whale stocks and thus make possible the orderly development of the whaling industry".[30] Accordingly, the IWC's original focus was not on developing an international whale protection regime but rather to establish a club for whaling nations and the industries they supported as a means to regulate stocks and promote a sustainable whaling industry.[31] Nonetheless, commentators have pointed to the duality of the IWC's purpose statement, and considerable ink has been spilled exploring whether, or to what extent, the IWC must balance conservation *and* utilization, or which should take priority. Lyster, for instance, has argued that whale conservation should, hierarchically, be prioritized over maintenance of the whaling industry because "conservation" is a prerequisite to sustained whaling.[32] This interpretation supports the view that the ICRW/IWC regime can modernize alongside cultural norms and social expectations.[33]

In contrast to the 1931 *Convention* and 1937 *Agreement*, the ICRW provided the IWC with greater power to regulate whaling through a Schedule which is open to amendment by a three-quarters voting majority. The ICRW Schedule sets forth regulations regarding (1) protected and unprotected species; (2) open and closed seasons; (3) open and closed waters, including designated sanctuary areas; (4) size limits for each species; (5) time, methods, and intensity of whaling; (6) types of gear allowed; and (7) methods of measurement and catch returns and other records. In addition, the Commission was meant to "encourage, coordinate, and fund whale research, [publish] the results of these and other scientific research, and promote studies into related matters, such as the humaneness of the killing operations".[34] An "opt-out" clause is fundamental to the IWC, such that whenever changes are made to Schedule provisions, there is a 90-day period before they become effective, during which any state can lodge an objection. An objection triggers a further 90-day period where other states can choose to join the objection; after which the amendment to the Schedule becomes binding on all states that have not objected.

Despite its status as a whaling nation, Japan did not attend ICRW negotiations due to post-WWII limits imposed on their sovereignty. Immediately following WWII's conclusion in the Pacific theater, Japan was placed under military occupation led by the United States, and Japanese fisheries were restricted to the state's coastal waters. Japan faced a dire food shortage because its fishing and meat industries were devastated during the final stages of the war. This food situation became a severe problem, and in 1946, the United States unilaterally authorized Japan to hunt whales in Antarctica and encouraged the retrofitting of tanker ships into whaling vessels as a partial response to the crisis.[35] Other IWC Contracting Governments opposed this decision and "the Japanese were seen by New Zealand and Australia not only as undesirable competitors, but were also accused of not having conformed with international agreements before the war".[36] In response to international pressure, the American State Department announced that Japan would comply with all the rules of the ICRW/IWC.[37] In the years directly following WWII, whale meat would constitute nearly half of Japan's meat supply[38] and would remain the cheapest form of animal protein until the mid-1950s.[39]

In 1951, American military occupation ended and Japan officially joined the IWC. Japan was not yet welcome to join the United Nations (UN) because of lingering post-WWII anxiety and mistrust.[40] Holm suggests that Japan's joining the IWC was politically strategic because it "enabled the country to restore its damaged international reputation and to present itself as a reliable partner complying with international standards".[41]

Each Contracting Government to the ICRW represents a unique social, political, and cultural perspective on whaling. These perspectives were primarily aligned in the 1950s during the early efforts to manage commercial whaling. Unfortunately, however, the catch limits in the ICRW Schedule were consistently set too high, despite observed species declines. Additionally, an overall quota system led to a "Whaling Olympics" with each Contracting Government racing to take the largest share of the total permitted catch. Such competitive—and unsustainable—behaviour was facilitated by the fact that the ICRW prohibits restricting "the number or nationality of factory ships or land stations [and cannot] allocate specific quotas to any factory ship or land station or to any group of factory ships or land stations".[42] Furthermore, Hoey observes that because the Schedule limited "the seasons in which whalers could hunt, nations worked to develop the most efficient ships and killing techniques in order to maximize their share of the whaling quotas" during the whaling season.[43]

As an attempt to resolve the 1950s catch limit process, in 1961, the IWC "appointed a special committee of three experts in population dynamics drawn from countries not engaged in pelagic Antarctic whaling. They were asked to make an independent analysis of the baleen whale stocks and appropriate recommendations to the commission".[44] This report resulted in another IWC deadlock as whaling countries could not accept the committee's advice to drastically reduce the 1963–1964 catch limit.

Environmental awareness continued to grow in the 1960s, and emerging environmental ethos within influential Contracting Governments spilled over into the ICRW/IWC. Simultaneously, scientific studies began to inquire about the intelligence of whales and other cetaceans,[45] and "a fear that the largest living creatures on the planet could be wiped out, led to a dominant anti-whaling discourse appearing in the US, Australia and Western Europe".[46]

Thus, the "preservationist" movement was born and would continue to grow within influential states and more broadly at the IWC itself. For example, the United States' delegation to the 1972 UN Conference on the Human Environment (UNCHE) in Stockholm successfully lobbied for, and ultimately secured, passage of a resolution calling for a 10-year moratorium on commercial whaling. Japan was taken off guard by this UNCHE declaration because Kumao Kaneko, the Japanese diplomat at UNCHE, assumed any discussion regarding whales and the future of commercial whaling would occur at the IWC.[47] Japan was steadfastly opposed to the emerging trend in cetacean protectionism and commentators such as Hirata suggest that Japan's sustainable use stance is multifaceted in that it was spurred by post-WWII nationalism and an emphasis on Japan's unique cultural experience with whale consumption (*gyoshoku bunka*).[48] Specifically, Japan's linguistic conceptualization of whales as a fish (the "symbol for whale (鯨—pronounced 'kujira') includes within it a component—魚—that means fish (pronounced 'sakana'").[49] Sumi emphasizes that "[f]or the Japanese people, the whale is not only a food source, but also a basis of culture".[50]

At the 1972 IWC's 24th Annual Meeting, which convened 10 days after UNCHE concluded, the United States proposed a 10-year moratorium on commercial whaling. The IWC, at this moment in time, was opposed to a moratorium and the proposal failed, with Contracting Governments voting as follows: four Contracting Governments in favour, six opposed, and four abstentions.[51] Renewed efforts to institute a moratorium one year later at IWC25 were also unsuccessful, with eight Contracting Governments in favour, five opposed, and one abstention.[52] Although the moratorium was not accepted, several conservationist policies were passed at IWC24, such as phasing out killing the Antarctic fin whales by 1976; and replacement of the "Blue Whale Unit" that was previously used to determine catch quotas across all whale species with species-specific catch quotas.

The environmental movement's influence continued to build momentum and legitimacy through the 1970s. Well-organized environmental groups such as Greenpeace and the World Wildlife Fund began to exert influence at the IWC by attending annual meetings as non-voting observers, circulating position papers, presenting research, and publicly protesting whaling states.[53] In 1975, the IWC changed course again by adopting Australia's proposed "New Management Procedure", whereby whale species were classified into three categories based on maximum sustainable yield calculations from the IWC's Scientific Committee, effectively creating a moratorium on some whale species and strictly limiting the allowable catch quotas for others. Japan, in turn, recruited many non-whaling countries to join the IWC to vote for Japanese interests in exchange for aid and development projects[54] and merged its major fishing companies into the "Japan Joint Whaling Company".[55]

International law's overarching approach to whaling would undergo arguably its most significant renovation to date in the early 1980s. External to the ICRW/IWC, ongoing negotiations for the UN Convention on the Law of the Sea at the third negotiating conference (UNCLOS III), 1973–1982 were finalizing and settling a number of critical oceans law issues.[56] Somewhat unexpectedly,[57] cetacean conservation provisions were negotiated during UNCLOS III and made their way onto the Informal Composite Negotiating Text (ICNT) that was voted on in 1982. The two relevant provisions of UNCLOS, namely Articles 65 and 120, provide as follows:

Article 65

Nothing in this Part restricts the right of a coastal State of the competence of an international organization, as appropriate, to prohibit, limit or regulate the exploitation of marine mammals more strictly than provided for in this Part. States shall cooperate with a view to the conservation of marine mammals and in the case of cetaceans shall in particular work through the appropriate international organizations for their conservation, management and study.

Article 120

Article 65 also applies to the conservation and management of marine mammals in the high seas.

While a detailed review of Article 65's content is beyond the scope of this chapter, a few salient features help explain how UNCLOS contributed to international cetacean management. First, Article 65 exempts all marine mammals from the goal of optimum utility that otherwise guides living marine resource management in the Exclusive Economic Zone (EEZ).[58] Second, Article 65 imposes a mandatory duty to cooperate and work through international bodies for cetacean conservation, management, and study. Article 120 extends this obligation to cover high seas cetacean management. While Article 65's language does not expressly advance a protectionist agenda, remarks made by U.S. Ambassador Elliot Richardson when these Articles were included in the ICNT are revealing. Specifically, he observed "[t]he new provision establishes a sound framework for the protection of whales and other marine mammals" through which "States and international organizations may pursue the future protection of these wonderful creatures for generations to come".[59] Pro-whaling states, including Norway and Iceland, supported including these articles in the ICNT, Japan offered the following commentary and its qualified support:

My delegation continues to consider that the concept of optimum utilization also applies to marine mammals. Consequently, there is no need to single out marine mammals in a special provision, or to focus on cetaceans in such a provision. As a practical matter, however, we can support this text on the understanding, with regard to the second sentence, that these activities do not necessarily need to be undertaken simultaneously with the first sentence, but on an individual (per species) basis when appropriate with consultations with other nations.[60]

Finally, it is notable that Article 65 references "international organizations" in the plural, which opens the door to alternative venues for cooperation besides the IWC.

As noted by Forkan, it was around this time that NGOs and the United States began to lobby in favour of a significant renovation at the IWC:

> The U.S. began to push for a re-negotiation of the ICRW to make it an International Cetacean Convention. The NGO community also strongly supported renegotiating the treaty calling for an International Cetacean Commission (ICC) – not only changing the emphasis from whaling to the whales themselves but to broaden jurisdiction to small cetaceans such as dolphins and porpoises.[61]

The Protectionist Agenda Takes Root and Phase II of the IWC (1982–Present)

The protectionist ethos that influenced the negotiation of Articles 65 and 120 of UNCLOS was now taking root at the IWC and began manifesting in tangible outcomes. The first significant development was the approval of the Indian Ocean Sanctuary in 1979. Article V(1)(c) of the ICRW explicitly allows the IWC to fix open and closed waters, including the designation of sanctuaries. As with all Schedule amendments, proposed sanctuary designations must be adopted by a three-quarters majority. The Indian Ocean Sanctuary received the necessary three-quarter majority vote, with 16 Contracting Governments voting in favour, 3 against (including Japan), and 3 abstaining. This Sanctuary covered the entire Indian Ocean to 55°S and prohibited all whaling for a 10-year period.

After the Indian Ocean Sanctuary's establishment, the IWC created a Working Group in 1981 to further develop the concept of a whale sanctuary and its dominant characteristics. While the IWC did not formally adopt the guidelines created by the Working Group, they have been frequently cited in subsequent sanctuary submissions and counterarguments.[62] The Working Group Report described that the primary sanctuary objective is to ensure the conservation and utilization of whale resources, this being consistent with the Preamble and Article V of the ICRW.[63]

By the early 1980s, IWC membership had increased to 39 Contracting Governments, and many new participating states, including the Netherlands and New Zealand, were opposed to commercial whaling.[64] Further, a coalition between powerful anti-whaling states, including the United States, Australia, New Zealand, Great Britain, and Germany, enabled effective control of more than half of the IWC votes and thus became positioned to influence several IWC decisions.[65] The most significant development occurred in 1982 when the IWC voted in favour of a temporary moratorium on commercial whaling. Despite Japan's efforts to oppose the moratorium, the vote surpassed the three-quarters majority and passed with 25 votes in favour to 7 against (with 5 abstentions).

The 1982 moratorium vote elicited emotional reactions from a number of delegations. Japan's formal response was to object to the moratorium under Article V(3) of the ICRW and to reiterate an intention to continue commercial whaling since they perceived "saw no legal or moral obligation to accept any decision of the commission".[66] Other pro-whaling states, including Peru, Norway, and the Soviet

Union, also objected, thereby legally exempting them from the moratorium.[67] The following paragraph was subsequently added to the Schedule:

> [C]atch limits for the killing for commercial purposes of whales from all stocks for the 1986 coastal and 1985/86 pelagic seasons and thereafter shall be zero. This provision will be kept under review, based upon the best scientific advice, and by 1990 at the latest the Commission will undertake a comprehensive assessment of the effects of this decision on whale stocks and consider modification of this provision and the establishment of other catch limits.[68]

Pre-commercial moratorium, IWC meetings were largely dominated by matters relating to commercial whaling, such as catch limits and humane killing methods, but a moratorium meant that the Commission could shift its management gaze in a different direction. During IWC36 (1984), the Commission established a panel that would meet before the next annual meeting to review the future operations of the IWC's Scientific Committee and to refocus its priorities during the moratorium. The Commission also adopted a US proposal to establish a working group to consider the future activities of the Commission itself and its continuing responsibilities. These initial steps signalled a subtle shift to the emergence of the protectionist agenda at the IWC.[69]

For example, in 1984, the Scientific Committee drew up a list of stocks known or believed to be adversely affected by whale watching or other human activities and took note of the 1983 world guide to whale watching, *The Whale Watchers Handbook*.[70] The Committee subsequently also presented information regarding threats to cetacean habitats—namely pollution in the form of "man-made noise and chemical toxicants".[71] Although this work did not result in any formal resolutions, the Committee highlighted the need to further study these new threats and intimated that potential regulatory measures could ultimately be warranted.[72]

When the moratorium took effect, the working group established in 1984 presented several areas the IWC could focus its work on, including: comprehensive stock assessments for populations being utilized under objection to the moratorium or special permit; monitoring of regulations other than catch limits; Aboriginal/subsistence whaling; sanctuaries; publications and statistics; special permits; and several other activities, particularly concerning research.[73] A formal motion was presented to review regulatory measures other than catch limits and to consider whether the intent of the Convention itself allowed for such measures. Further, the IWC convened a workshop on bycatch of small cetaceans in gillnet fisheries, which resulted in the determination that this issue was beyond the IWC's regulatory jurisdiction better suited for discussion at the UN Environment Program (UNEP).[74] Still, and on the Scientific Committee's recommendation, the Commission endorsed further studies on the effects of pollutants on cetaceans.[75]

Although Japan was not initially legally bound by the 1982 moratorium and continued to commercially whale, the United States threatened to enforce the 1979 Packwood-Magnuson Amendment, which would immediately disrupt Japan's fishing quota in the United States' EEZ if Japan continued its whaling operations.[76] To maintain international relations, Japan officially abandoned commercial whaling in 1987 and, instead, initiated a scientific whaling program pursuant to Article VIII of the ICRW. This article allows any Contracting Government to issue to its nationals a special permit "to kill, take and treat whales for purposes of scientific research" provided that the IWC is notified.[77] Japan accordingly notified the IWC of the "Japanese Whale Research Program under Special Permit in the Antarctic" (JARPA), where they would hunt approximately 300 minke whales annually for scientific purposes. The IWC notified Japan of its non-binding objection to their proposal, but the IWC lacked the authority to formally sanction Japan. Japanese whaling ships entered Antarctica in 1988 and political international relations faltered as "the USA saw this as a breach of the moratorium and banned all Japanese fishing activities in the American EEZ".[78] The era of special permit scientific whaling was underway.

Some commenters have noted surprise with Japan's decision to pursue a scientific whaling campaign rather than simply withdrawing from the IWC. Epstein offers one potential explanation—post-WWII, Japan had worked diligently to build a reputation as an advocate of international standards, cooperation, and international law and did not want to diminish this perception.[79] Holm notes that Japan wanted to continue to hunt whales outside of its EEZ, specifically within the 50 million square kilometre Southern Ocean Whale Sanctuary, and this opportunity would have been lost if it left the IWC, owing to its ratification of UNCLOS.[80] Japan also hoped that through scientific whaling, it would soon be able to provide the scientific data on the status of whale stocks that would lead to the IWC reversing the moratorium.

The USSR emerged as a strong voice in favour of amending the ICRW in view of changing circumstances and the "norms of the International Law of the Sea".[81] In 1988, at IWC39, the USSR formally proposed to amend the ICRW on the basis that the cessation of commercial whaling had eliminated the IWC's traditional regulatory function and brought "research and conservation objectives to the forefront".[82] Other Contracting Governments, such as New Zealand, the United Kingdom, and Sweden, argued that this amendment was unnecessary since the Commission's evolving work fell within the ICRW's Convention and that formal amendments were not a priority.[83] In view of the differences of opinion from many Contracting Governments (e.g., with respect to issues of the IWC's jurisdiction over small cetaceans, the scientific permits for whale harvesting, and the effect of UNCLOS) over the interpretation of the 1946 International Convention on the Regulation of Whaling, the Commission agreed to "establish a working group charged with the responsibility of examining questions related to the operation of the 1946 Convention".[84] In 1989 at IWC40, the USSR submitted a draft Resolution that focused on whale conservation and research, arguing the "flexible and adaptable nature of the present Convention cannot be extended beyond the original intentions of the treaty, and must now be seen in context of the Convention of the Law of the Sea".[85] Japan opposed this motion and argued that the objectives of the Convention were to conserve whale stocks for their rational utilization and to develop the whaling industry.

Japan opposed altering the ICRW and expressed concern that other Contracting Governments were focused exclusively on conserving whale stocks without allowing for their utilization. This contest ultimately concluded with an impasse.

In 1992, some 13 years after the Indian Ocean Sanctuary's designation, France proposed the Southern Ocean Sanctuary, which was voted into force in 1994. Similar to the Indian Ocean Sanctuary, commercial whaling was prohibited in the Southern Ocean Sanctuary. In moving this proposal, France offered the following rationale: the primary purpose of this sanctuary is to

> contribute to the rehabilitation of the Antarctic marine ecosystem by reinforcing and complementing other measures for the conservation of whales and the regulation of whaling, in particular by the protection of all Southern Hemisphere species and population of baleen whales and the Sperm whales on the feeding grounds.
>
> (IWC/44/19)[86]

Japan formally objected to the Southern Ocean Sanctuary under Art. V(3), but only to the extent that it applied to harvesting minke whales, thereby ostensibly accepting the Southern Ocean Sanctuary and its restrictions.[87]

Japan did, however, continue to protest against the IWC's use of whale sanctuaries, relying on Burke's legal opinion that "the sanctuary is irrelevant and unjustified if other regulations already prohibit any take of particular species, which is the actual situation. In this actual situation, the Southern Ocean Sanctuary is redundant and has no place".[88] The IWC's use of whale sanctuaries was also susceptible to the criticism that this approach was largely "symbolic" since it only served to protect whales "from one of the many threats endangering their survival" and also ambiguous in that the "Whaling Convention does not provide a precise definition of the scientific basis for the establishment of a closed area, thus leaving undetermined the kind of evidence that needs to be brought forward by proposing member states".[89]

At IWC44 in 1993, New Zealand successfully advanced a new protectionist policy by proposing a Resolution on Small Cetaceans, which passed.[90] Japan opposed this Resolution, arguing that the IWC did not have the necessary competence to ban the catch of small cetaceans, but the resolution passed with 15 votes in favour, 5 against, and 7 abstentions.[91] Accordingly, the Scientific Committee established a regular agenda item relevant to the long-term management and conservation of small cetaceans and further committed to addressing the impact of "environmental changes on whale stocks, [...] and that it should develop practical means to address the [concerns]".[92]

The momentum of the 1993 meeting continued the next year at IWC45 where a significant shift in the IWC's ability to pass meaningful protectionist policies occurred. The United Kingdom submitted a paper explaining how the public perception of cetaceans had changed in the last decade and evidence that the whale watching industry was growing by roughly 49% each year.[93] The United Kingdom advocated that "the IWC can play an important role by collecting and disseminating information, providing advice about the use of whale watching for cetacean research, and preparing guidelines to set high standards of practice for the industry".[94] Japan expressed concerns with the proposal, specifically with applying the term "sustainable use" to whale watching. Instead, Japan argued that whale watching was outside of the purview of the IWC, and that the IWC should return its focus to sustainable cetacean harvesting. Japan further suggested that continued research into whale watching could cause harm to whale stocks.[95] Ultimately, the Commission noted Japan's concerns but agreed to establish a working group to bring forward a policy on whale watching at a future meeting.[96] Over the next several years, the Scientific Committee researched the impact whale watching had on cetaceans, which cumulated in a set of guidelines presented at IWC47 that were designed to assist coastal states in managing the burgeoning industry.

At IWC47, the Scientific Committee also presented an update on its work researching how environmental changes were affecting whale stocks. Specifically, the Committee decided it should focus on two types of environmental changes: the first being climate change where the long-term impact was at an early stage, and second being pollution, indirect effects of fishing mortality, and habitat loss from shipping.[97] A resolution was passed to have the Scientific Committee prioritize research on the effects of environmental changes on cetaceans; and although Japan voiced concerns that the Convention for the Conservation of Antarctic Marine Living Resources (CCAMLR) already dealt with Antarctica and the Scientific Committee could become overworked by such a broad resolution, it ultimately supported it.[98] "Environmental concerns" have remained standing item on the IWC's Annual Meeting agenda and is an umbrella term for many environmental threats to cetacean survival. Some of the environmental concerns discussed by the Committee are underwater noise pollution from shipping and seismic surveys, chemical pollution and marine debris, habitat degradation, and climate change.[99] The IWC's focus in the area of environmental concerns has been to evaluate and report on environmental threats and suggest mitigation measures that could be employed by coastal States.[100]

During IWC48 in 1997, the Scientific Committee presented its study, "Objectives and Principles for Managing Whale Watching". The Commission passed a Resolution where coastal states should be made aware of these recommendations but noted that they were not binding and it was up to the individual regional governments to decide their policies.[101] The following year, at IWC50, the Scientific Committee presented an extensive report on the environmental threats to cetacean survival, specifically pollution, climate change and habitat, noise, fisheries, and other mortality events. Notably, Japan argued that continued research could be more effective via lethal means than the current non-lethal studies, but the IWC adopted a Resolution on Environmental Changes and Cetaceans, noting Japan's reservations, which provided that the IWC:

AGREES to establish a regular Commission Agenda Item entitled 'Environmental Concerns' under which the Scientific Committee would continue to report annually on its progress in non-lethal research on environmental concerns, and Contracting Governments could report annually on national and regional efforts to monitor and address the impacts of environmental change on cetaceans and other marine mammals.[102]

Key policies introduced during this time focused on whale protections, such as methods of mitigating bycatch. Bycatch, which is the accidental capture of a non-target species by fishing gear, kills approximately 300,000 cetaceans per year and has been an area of concern and cause for study by the Scientific Committee since the 1980s.[103] In 1997, the IWC passed a Resolution on Bycatch Reporting and Bycatch Reduction, which formally called on all Contracting Governments to improve their monitoring and reporting of all cetaceans taken accidentally by fishing gear, and to begin to report those accidental catches at all future IWC meetings.[104]

In 2003, a new Conservation Committee was proposed that would include all parties to the IWC, but the IWC Chair aptly noted it would not solve the problems within the IWC[105] "given that neither side was willing to seriously engage with the other", and so discussions about either a return to commercial whaling or further drift towards more holistic whale protection and conservation would not result in solutions.[106]

At IWC58 in 2006, pro-sustainable use Contracting Governments took direct aims at the commercial moratorium with the St. Kitts and Nevis Declaration, which stated "that the moratorium is no longer necessary for conservation purposes in light of abundance of certain whale stocks such as the minke whale".[107] This declaration proposed the normalization of the IWC's activities with a return to commercial whaling. This contentious declaration passed with a vote of 33 in favour, 32 against, and 1 abstention, but it did not have legal effect and thus did not change the status of the commercial whaling prohibition.

In a continued push for a return to managed commercial whaling, Japan hosted an unofficial "Conference for the Normalization of the IWC" the following year, but whale preservationist states declined to attend and the IWC did not accept the meeting report into its archive. At the same time, an official Steering Group was created to "bridge the divide within the IWC and rebuild trust in the organization".[108] The Steering Group created and presented a "draft Consensus Decision to Improve the Conservation of Whales" in 2010 but simultaneously acknowledged "that no state was changing their negotiating stance and that whaling issues would continue to be the province of the IWC".[109] It is unsurprising that "[d]espite over 30 meetings in two days a compromise between preservationist and conservationist states could not be agreed".[110]

Japan's initial scientific whaling JARPA program ran from 1987 to 2005, and then its "second phase", called JARPA II, commenced in 2005. The JARPA II research plan "provided for an annual catch of 50 fin and humpback whales and upwards of 850 Antarctic minke whales—over double the numbers of its predecessor JARPA, which resulted in the killing of more than 6,700 minke whales in the span of 18 years".[111] Clapham notes "this transition occurred before the IWC's Scientific Committee had been given a chance to review the final results of JARPA I".[112]

Although Japan kept whaling under the auspices of scientific whaling, the prospect of their ultimate goal of having the commercial moratorium lifted continued to dwindle. As Holm opines, "[m]aintaining the status quo was acceptable to many IWC anti-whaling member states as it effectively froze the whaling conflict and made commercial whaling almost impossible".[113] Joji Morishita, the former head of the Japanese delegation to the IWC, explained Japan's long-term strategy as follows:

> If it were only about whaling issues, [maintaining the status quo] would be acceptable. I admit that the whaling industry is only a small part of Japanese interests. If this were our only interest, the time and energy we spend on it would probably be too great. But if the same logic were extended to other areas, [whaling] would set a precedent.[114]

On May 31, 2010, the Australian government filed formal proceedings against Japan at the ICJ. Australia's application alleged Japan violated the following: its ICRW obligations under Article VIII to conduct a program "for purposes of scientific research", of paragraph 10(e) of the Schedule to observe in good faith the zero catch limit for commercial whaling, and of paragraph 7(b) of the Schedule to act in good faith to refrain from undertaking commercial whaling of humpback and fin whales in the Southern Ocean Sanctuary.[115] The ICJ's judgment concluded that JARPA II's lethal sampling program failed to constitute whaling "for the purposes of scientific research" pursuant to Article VIII of the ICRW and accordingly ordered the cessation of these activities. In reaching its decision, the Court considered: the scale of the lethal research employed, noting that JARPA and JARPA II had roughly identical objectives but vastly increasing lethal sampling quotas; the manner in which sample sizes were selected and a disparity of intended sample sizes with the number of whales actually killed; JARPA II's unlimited time frame; and that only two peer-reviewed papers had been published since 2005.[116]

Beyond ordering JARPA II's immediate cessation, the ICJ ruling had implications for the IWC's internal struggles. As Clapham notes, "the ICJ judgement represented the first time an objective international body, independent of the IWC, had reviewed and assessed Japan's scientific whaling—and in doing so had found it wanting".[117] The decision also pointed to significant organizational faults within the IWC's Scientific Committee, identifying the extent to which scientific discussions were politically motivated and biased—Japan, after submitting a whaling proposal, was involved in both reviewing and writing the resulting scientific reports. Further, the ICJ emphasized the history of conflict at the IWC's foundation. The Court interpreted Article VIII with the ICRW's dual objectives in mind, noting the tension within the IWC's twin objectives of both whale protection and sustainable exploitation of whale stocks; with neither objective occupying formal priority.[118]

Following the ICJ's decision, Noriyuki Shikata, as spokesperson for Japan, said Japan was "disappointed" by the decision but would abide by it "as a state that places great importance on the international legal order and the rule of law as a basis of the international community".[119] In November 2014, Japan announced a new plan for scientific whaling in the Antarctic that would commence in 2015. The new programme titled the "New Scientific Whale Research Program in the Antarctic

Ocean" (NEWREP-A). When the proposal was submitted to the Scientific Committee, the review panel found that "it was not able to determine whether lethal sampling is necessary to achieve the [program's] two major objectives", mainly because "the proposal contained insufficient information for the panel to complete a full review".[120] Brierley, a marine ecologist at the University of St. Andrews and member of the IWC's scientific panel, disclosed that other panellists warned him at the beginning of the process that Japanese whaling was "inevitable" and that "their review was a waste of time".[121] As Telesketsky, states

> [t]here is nothing specific in either Article VIII or in the Schedule to prevent a State from ultimately issuing a special permit that the Scientific Committee may have reservations about as long as the Committee has been given adequate opportunity to 'review and comment'.[122]

As the impact—and legality—of NEWREP-A was considered, the IWC continued to move forward with protectionist measures. For example, in addition to the Bycatch Mitigation Initiative and continued work in the area of whale watching, explored above, in 2016 the IWC established the Strandings Initiative following a multi-disciplinary expert workshop meant to "assist the IWC in its efforts to build global capacity for effective cetacean stranding response and promote the IWC as a leading body for the provision of advice through the development of practical guidance for responders".[123] The Initiative is overseen by the Commission's Scientific and Welfare Committees and is currently developing two training programs. The first is to focus on emergency response for live standings, and the second is focused on necropsy work for large-scale mass mortality events.[124] Although the research and materials produced by this Initiative are available to all coastal nations, the training program is only accessible to Contracting Governments.[125]

At IWC67 in 2018, Japan presented a "Way Forward" proposal that Japanese Vice-Minister for Foreign Affairs, Mitsunari Okamoto, stated was "mindful of the interests of all [IWC] Contracting Parties".[126] The two main reforms proposed were to establish a Sustainable Whaling Committee to recommend commercial catch limits for abundant whale populations, and second, an amendment to the Convention to change the voting mechanism from a three-fourths majority to a simple majority for amendments to the Convention Schedule.[127] Precisely, 17 Contracting Governments supported the proposal, but the vote would have to be unanimous for it to pass owing to the proposed treaty amendment.

After Japan's "Way Forward" proposal was defeated, Japan acted on its longstanding threat and announced its withdrawal from the IWC on December 26, 2018, effective June 30, 2019. Chief Cabinet Secretary, Mr. Suga announced:

> Although scientific evidence has confirmed that certain whale species/stocks of whales are abundant, those member States that focus exclusively on the protection of whales, refused to agree to take any tangible step towards

reaching a common position that would ensure the sustainable management of whale resources.[128]

Japan's withdrawal from the IWC does not mean Japan can engage in whaling free from any legal constraints in the high seas.[129] To comply with the Convention on International Trade in Endangered Species of Wild Fauna and Flora (CITES),[130] Japan must only commercially whale within its EEZ to avoid international trade in whale meat.[131] For instance, in October 2018, the CITES Standing Committee and Secretariat declared that Japan's scientific whaling of endangered sei whales in the North Pacific violated Article III, paragraph 5(c) of CITES. The Standing Committee determined that because sei whales were hunted beyond Japan's EEZ, brought into Japan, and then sold for commercial purposes, it constituted illegal international trade.[132] Japan also remains a Contracting Party to UNCLOS, and, as discussed above, Article 65 of UNCLOS compels Japan to cooperate with other states and to work through the appropriate international organizations for the conservation of marine mammals and "in the case of cetaceans shall in particular work through the appropriate international organizations for their conservation, management and study".[133] Thus far, Japan's EEZ whaling has been conducted in accordance with the catch limits recommended by the IWC and has otherwise been prosecuted in agreement with international law. The question that remains, however, is what Japan's exit from the IWC means for this regime's functionality going forward.

Another Door Opens?

The narrative presented in the preceding part demonstrates the IWC's strained transformation through two phases of existence; a transformation that has seen it evolve from a whaling management organization that functioned by overseeing the conservation of whale for the purposes of commercial whaling to one that shepherds a ban on commercial whaling whilst gradually developing additional whale protection measures. This progression has been circuitous and gradual, influenced by the push-pull dynamics that exist between the IWC's Contracting Governments. As stated by then IWC Chair and Vice Chair: "very different views exist among the members regarding whales and whaling and that this difference had come to dominate the time and resources of the commission at the expense of effective whale conservation and management".[134] Japan's withdrawal from the IWC would seem, at least superficially, to open the door to a stronger protectionist agenda and present an opportunity to "focus on other conservation efforts which the pro- and anti-whaling debate has often distracted from"[135]; in other words, the actualization of Phase III of the Commission existence. This portion of the chapter considers options in this regard and weighs the impact of Japan's withdrawal on their prospect of success.

Options for Enhanced Protectionist Action

In our opinion, the most logical and direct route to a robust protectionist agenda at the IWC is through resolutions and/or Schedule amendments aimed at progressive and meaningful action on contemporary threats to the long-term survival of whale populations. Achieving this, however, is limited by the structure of the ICRW/IWC regime itself.

To date, the majority of progressive efforts taken in response to contemporary threats to whales, including on issues such as bycatch and entanglement, environmental degradation, and climate change, have largely followed the same trajectory: the IWC has committed to studying these issues, sought to encourage greater international cooperation (either within the IWC or with other international legal regimes) and has encouraged Contracting Governments to address these issues more effectively through municipal law and policy. The exception to this approach is the designation of Whale Sanctuaries, which has required formal amendment of the IWC's regulatory Schedule. The reality is, however, the opportunity for more aggressive protectionist management measures remains inherently limited by the ICRW's structure.

The treaty itself prescribes the IWC's regulatory functions, which are predominately directed towards commercial whaling and sustainable use (e.g., catch quotas, open and closed seasons, hunting methods). A second limiting factor is the ICRW's object and purpose—to "ensure proper and effective conservation and development of whale stocks and thus make possible the orderly development of the whaling industry".[136] While formal amendment of this purpose statement would require unanimous consent, the ICRW/IWC's history clearly demonstrates that its management priorities can shift with changing membership and shifting attitudes. The ICJ, in *Whaling in the Antarctic*, opined that the very structure of the ICRW and its Schedule, understood in the context of the IWC's functionality, has "made the Convention an evolving instrument".[137] While this perspective on the IWC's continued evolution has not gone unchallenged,[138] certain developments suggest that the IWC has been drifting towards a more holistic perspective on whale management that emphasizes the important role of whales within ocean ecosystems.

At IWC66 in 2016, the Contracting Parties passed the Resolution on Cetaceans and Their Contributions to Ecosystem Functioning[139] with 36 Contracting Governments in favour, 16 against (including Japan and other pro-sustainable use Contracting Governments), and 9 abstentions. The Resolution "recognized the need to include consideration of the contributions made by live cetaceans and carcasses present in the ocean to marine ecosystem functioning in conservation, management strategies, and decision making" and also directed the IWC to cooperate with other international organizations working on this issue.[140] To this end, the IWC convened a workshop in 2017 focussed on cetacean nutrient transport both in the water column and between whale foraging and breeding grounds, the impact of whale falls on biodiversity and carbon sequestration, and the role of cetaceans as predators and prey.[141]

At IWC67 in 2018, the IWC adopted the non-binding Florianópolis Declaration, which articulates that the role of the IWC in the 21st Century "includes *inter alia* its responsibility to ensure the recovery of cetacean populations to their pre-industrial levels", reaffirmed "the importance in maintaining the moratorium on commercial whaling", and committed to deeper cooperation and coordination with the "Convention on Biological Diversity, the Convention on the Conservation of Migratory Species of Wild Animal, the Convention on the Conservation of Antarctic Marine Living Resources and the World Tourism Organization".[142] This has been followed with a number of working groups and reports that have affirmed the critical role that cetaceans occupy in the marine environment and their socio-economic value. The shift, then, towards—and possibly into—Phase III, is ongoing and it is prudent to weigh the impact of Japan's withdrawal in potentially hastening this transition.

Measuring the Effect of Japan's Withdrawal from the IWC

Japan may only represent one vote amongst many at the IWC, but Japan's influence extends beyond its vote share since it has been one of, if not the most, vocal supporters of the IWC's return to a commercial whaling organization. As noted in the first part, new protectionist measures have usually been met with vocal opposition from the Japanese delegation, which has sought to re-focus the IWC's management efforts on the sustainable utilization of certain whale stocks. The impact of Japan's withdrawal may also be amplified by the fact that Japan's international approach to whaling politics has extended to vote buying: "exchanges of valuable consideration for a vote or decision on a matter concerning the wider community of states".[143] Gillespie's investigation of the long history of IWC vote buying, with a particular focus on Japan's influence on small Pacific Island states, notes that the practice was so prominent that it warranted an IWC resolution in 2001 affirming the importance of the international principle of good faith as follows:

> The complete independence of sovereign countries to decide their own policies and freely participate in the IWC (and other forums) without undue interference or coercion from other sovereign countries.[144]

A well-publicized 2010 investigation by the London newspaper *The Sunday Times* worked to daylight Japan's vote buying efforts at the IWC. Posing as Swiss philanthropists, two undercover reporters offered millions of dollars in aid to developing states if they would vote against whaling quotas.[145]

Dippel's assessment of vote buying at the IWC examined voting records from 1991 to 2006, a period during which IWC membership swelled from 40 to 70 countries. Dippel's data analysis points to Japan increasing its spending on foreign aid to countries receiving that voted in favour of Japan's position.[146] The insinuation being that substantial foreign aid packages have been used to persuade small countries, often with no whaling tradition or coastline, to join the IWC, and vote in favour of Japan's position.[147] Dippel's analysis also examined the timing of Japan's foreign aid spending and found that aid was provided only after countries voted with Japan's interests and then increased again the following year.[148] This review concluded in ended 2006, which Dippel justified on the basis that "[v]oting on any contentious issues effectively stopped altogether in 2007".[149] However, one contentious voting issue that has recently been stymied by the pro-sustainable use voting block is the proposal for additional whale sanctuaries.

A Southern Atlantic Whale Sanctuary was first proposed by Brazil in 2001; and in 2002, a South Pacific Whale Sanctuary was proposed by New Zealand. Both have been considered on multiple occasions. In 2011, Japan and 20 other pro-sustainable use Contracting Governments walked out of the IWC meeting, thereby thwarting a vote on a Southern Atlantic Whale Sanctuary, which forced the matter to be tabled owing to the lack of quorum.[150] In 2018, 11 of the states that had supported the Japanese "Way Forward" proposal voted against the creation of the South Atlantic Whale Sanctuary, and it was narrowly defeated with 39 Contracting Governments in favour, 25 opposed, and 3 abstentions. The IWC acknowledges this contentious record "an additional proposal for a Sanctuary in the South Atlantic Ocean has been repeatedly submitted to the Commission [...but] has not achieved the three-quarters majority of votes needed".[151]

Tempering Expectations—IWC68

Owing to Covid-19 pandemic delays, there was a four-year break between IWC67 and IWC68, which convened in October 2022. This was the first plenary IWC meeting since Japan's withdrawal. No other Contracting Governments withdrew from the IWC in the intervening years, putting to rest some of the concerns that the possibility of the IWC moving further in the preservationist direction, would trigger additional withdrawals.[152] Japan attended the meeting as a formal observer and offered the following statement at the meeting's commencement:

> Despite its withdrawal, Japan has not changed its basic policy that it is committed to international cooperation for the proper management of aquatic living resources. Japan has always been working for the proper conservation and management of whale resources based on scientific evidence, in coordination with international organizations, and it will continue to contribute toward this goal. With respect to the IWC, Japan will continue its cooperation, especially in the Scientific Committee [...].
>
> Also, Japan would like to emphasize that we maintain its unshakable faith that whale resources should be used sustainably based on the scientific evidence and this is consistent with the objectives of the ICRW. Japan will continue to pursue the science-based sustainable use of whale resources under the proper management, hoping that it will become the common goal of the IWC.[153]

The meeting featured the same themes that we have explored in this chapter. For example, Antigua and Barbuda submitted a draft resolution titled "Draft Resolution on the Implementation of a Conservation and Management Program for Whale Stocks aimed towards the lifting of the moratorium and the orderly development of the whaling industry".[154] As the title suggests, this draft resolution proposed to move the IWC in the direction of a "more balanced approach in consideration of its mandate".[155] Antigua and Barbuda, which does not whale but has close relations with Japan and has admitted to accepting financial aid and fishing infrastructure from Japan,[156] subsequently denied a desire to overturn the moratorium and tabled the resolution for reconsideration at IWC69.[157] Antigua and Barbuda was involved with a second resolution, this time as a joint-sponsor with Gambia, Cambodia, and the Republic of Guinea. This resolution, titled "Draft Resolution on Food Security", urged the IWC to consider nutritional and cultural needs in the context of whaling, and Aboriginal Subsistence Whaling in particular, and to enhance its cooperative efforts with the UN Food and Agricultural Organization. This resolution was also tabled for reconsideration at IWC69.

The one resolution that did pass was a "Resolution on Marine Plastic Pollution", as submitted by the Czech Republic on behalf of European Union member states.[158] This resolution, as is standard course for the IWC's contribution to global environmental problems that impact whales, commended work that was being by the UN Environment Assembly, which passed a resolution giving UNEP's Executive Director the mandate to convene an Intergovernmental Negotiating Committee to commence negotiations on a multi-lateral agreement to manage and limit plastic pollution through its entire lifecycle.[159] The resolution also directed the IWC to engage in the negotiations, as appropriate, and to continue the ongoing work that IWC Committees have been doing to study and consider methods of limiting the effect of plastic pollution on whales.[160]

The one proposed Schedule amendment was the creation of the South Atlantic Whale Sanctuary. A vote on this issue, which would require the three-quarters majority to pass, was never taken, however, since 16 Contracting Governments did not attend the scheduled vote, which brought proceedings to a standstill owing to a lack of quorum.

In many ways, IWC68 is nothing more than a predictable next-step in the IWC's progression. There is no appetite to return to a commercial whaling organization but the Commission's contribution to a strong protectionist agenda remains stymied, even in Japan's absence, by a vocal minority and an inherently limiting enabling treaty. Thus, while Japan's exit from the IWC is still relatively fresh and it may be too early to draw definitive conclusions on the impact that this may ultimately have, it is clear to us that this development has not opened the door to a stronger protectionist agenda so much as it has nudged it slightly ajar.

Conclusion

Peters, in considering the impact of Japan's IWC withdrawal from the IWC, justifiably questions the utility of the IWC going-forward by asking "what then is the use of a whaling treaty that binds only those states which do not hunt whales anyway" and observes that with "Japan's exit, the Whaling Convention has rather become akin to other international treaty regimes which unite the 'good' ones under their umbrella, with potential violators remaining outside".[161] In our estimation, Japan's withdrawal from the ICRW does not signify either the IWC's demise or its sudden transition to a strong protectionist organization. Rather, it is another incremental development in the ICRW/IWC's continued evolution.

For better or for worse, the ICRW remains tethered to its original purposes and the IWC's regulatory function is bound by the strictures of its enabling treaty. While there is undeniably some leeway within these parameters to reflect the changing socio-cultural priorities of the Contracting Governments, the IWC—short of radical re-invention—is not likely to ever become an organization capable of effectively stewarding the world's great whale species. At best, the IWC has been able to finance research into whale threats and, more recently, to lobby for cetacean conservation interests in other international legal forums. The IWC has not, however, been able to make any meaningful and legally binding regulations based on the results of its work. While the IWC's recent recognition that whales are a critical part of ocean ecosystems that provide a range of valuable ecosystem services is a necessary development that demonstrates progress in the IWC's philosophy, this shift is not significant enough to declare that the IWC is now equipped to effectively steward international whale conservation through the 21st century. Without a radical re-imagining of its core function, achieved through compromise between Contracting Governments and amendment to the ICRW, the IWC seems destined to fade out of relevance.

Notes

1 Japanese Whaling Association, 'History of whaling' < http://www.whaling.jp/english/history.html > accessed 11 May 2023.
2 Fynn Holm, 'Japan's whaling policy: The reasons for leaving the International Whaling Commission' (2019) < https://www.researchgate.net/publication/337387724_Japan's_Whaling_Policy_The_Reasons_for_Leaving_the_International_Whaling_Commission>, 3. This is the English translation of Fynn's chapter, cited with permission, which is found in the following collection: Fynn Holm, 'Japan's Walfangpolitik: Die Gründe für den Austritt aus der Internationalen Walfangkommission', in David Chiavacci und Iris Wieczorek (eds.), *Japan 2019: Politik, Wirtschaft, Gesellschaft* (Iudicium, 2019) 126–151.
3 Ibid., 4.
4 Ibid.; J N Tønnessen and A O Johnsen, *The history of modern whaling* (Australian National University Press 1982) 142.
5 International Convention for the Regulation of Whaling 1946, 161 *UNTS* 72.
6 Ibid., last preambular paragraph.
7 Adrienne M Ruffle, 'Resurrecting the International Whaling Commission: Suggestions to strengthen the conservation effort', (2002), 27 *BJIL* 639, 647.
8 James E Scarff, 'The International Management of whales, dolphins, and porpoises: An interdisciplinary assessment', (1997), 6 *ELQ* 2 323, 349.
9 David S Ardia, 'Does the emperor have no clothes? Enforcement of international law protecting the marine environment', (1998), 19 *MIL* 497.
10 IWC, 51st Report (1999), section 7.1.3 of the Chairman's Report, online: <archive.iwc.int/pages/search.php?search=%21collection49&k=>; IWC, 52nd Report (2000), section 7.3 of the Chairman's Report, online: < archive.iwc.int/pages/search.php?search=%21collection49&k=>; IWC, 53rd Report (2001), section 7.4.1 of the Chairman's Report, online:<archive.iwc.int/pages/search.php?search=%21collection49&k=>;

Elisa Morgera, 'Whale sanctuaries: An evolving concept within the International Whaling Commission', (2004), 35 *ODIL* 319.

11 Canada was the only country to formally withdraw its IWC membership around the time of the whaling moratorium. Iceland withdrew its IWC membership in 1992 but rejoined in 2002 on the condition that it would not be bound by the moratorium, causing some countries such as the United States to not recognize Iceland's renewed membership. See also Gillespie, 2003; Iceland Rejoining, 2002.

12 Kieran Mulvaney, 'The killing goes on', in *The Guardian* (London, 12 June 1987) 20.

13 International Convention for the Regulation of Whaling 1946, Schedule, section 10(e).

14 Judgment, ICJ GL No 148, ICGJ 471 (ICJ 2014).

15 The Way Forward of the IWC: IWC Reform Proposal including a draft Resolution and proposed Schedule Amendment, IWC/67/08 (2018). Notably, Iceland and Norway, which both allow commercial whaling within their Exclusive Economic Zones, voted in favour of commercial whaling while Russia and South Korea abstained.

16 Niall Alexander Rand, 'Reforming the International Whaling Commission: Indigenous peoples, the Canadian problem and the road ahead', (2017), 19 *ICLR* 324.

17 Cameron S G Jefferies, 'International Whale Conservation in a changing climate: The ecosystem approach, marine protected areas, and the International Whaling Commission', (2018), 21 *JIWLP* 4, 239; Jeffrey J Smith, 'Evolving to conservation?: The International Court's decision in the Australia/Japan whaling case', (2014), 45 *ODIL* 301.

18 C S G Jefferies, *Marine Mammal Conservation and the Law of the Sea* (OUP 2016).

19 Michael Bowman, *Lyster's International Wildlife Law* (2nd ed., CUP 2010) 59.

20 Larry Leonard, 'Recent negotiations toward the International Regulation of Whaling', (1941), 35 *AJIL* 1, 90.

21 155 L.N.T.S. 349.

22 Ray Gambell, 'International Management of Whales and Whaling: An historical review of the regulation of commercial and aboriginal subsistence whaling', (1993), 46 The Arctic Institute of North America 2, 97.

23 David Day, *The Whale War* (Routledge, 1987), 129.

24 Gambell, 'International Management', 98.

25 Kazuo Sumi, 'The Whale War between Japan and the United States: Problems and prospects', (1989), 17 Dev J Int'l La & Pol'y 317, 350.

26 Ibid.

27 Ibid., 351.

28 Gambell, 'International Management', 98.

29 The original 15 IWC signatories were: Argentina, Australia, Brazil, Canada, Denmark, France, Iceland, Mexico, the Netherlands, Norway, Panama, South Africa, the Soviet Union, the United Kingdom, and the United States.

30 ICRW, last preambular paragraph.

31 Michal Kolmaš, 'International pressure and Japanese withdrawal from the Internaional Whaling Commission: When shaming fails', (2021), 75 *AJIA* 197, 202.

32 Simon Lyster, *International Wildlife Law* (CUP 1985) 20. For a full interpretation, see: Michael Bowman, "Normalizing" the International Convention for the Regulation of Whaling', (2008), 29 *MJIL* 293.

33 Bowman, 461.

34 Gambell, 'International Management', 99.

35 Kazuo Sumi, 'The Whale War', 350.

36 Holm, 'Japan's Whaling Policy', 5.

37 Ibid.

38 Kathy Glass and Kristen Englund, 'Why the Japanese are so stubborn about whaling', (1989), 32 *Oceanus* 45, 46.

39 Kazuo Sumi, 'The Whale War', 351.

40 Sadako Ogata, 'The changing role of Japan in the United Nations', (1983), 37 *JIA* 1, 29.

41 Holm, 'Japan's Whaling Policy', 5.

42 Gambell, 'International Management', 99.
43 Laura Hoey, 'The battle over scientific whaling: A new proposal to stop Japan's lethal research and reform the International Whaling Commission', (2017), 41 *WMELPRR* 435, 440.
44 Gambell, 'International Management', 99.
45 Karen Pryor, Richard Haag and Joseph O'Reilly, 'The creative porpoise: Training for Novel behavior', (1969), 12 JEAB 4, 653; Edward I Griffin and Donald G Goldsberry, 'Notes on the capture, care and feeding of the killer whale Orcinus orca at Seattle aquarium', (1968), 8 *IZY* 1, 206; John Cunningham Lilly, *The mind of the Dolphin: A nonhuman intelligence* (1st ed., Doubleday 1967); David Brown, 'Behavior of a captive pacific pilot whale', (1960), 41 *JM* 3, 342.
46 Holm, 'Japan's Whaling Policy', 6.
47 Maseru Nishikawa, 'The origin of the U.S.–Japan dispute over the whaling moratorium', (2020), 44 *Diplomatic History* 315, 327.
48 Keiko Hirata, 'Why Japan supports whaling', (2005), 8 *JIWLP* 129, 141.
49 Ibid.
50 Sumi, 'The Whale War', 318.
51 Hoey, 'The Battle', 441.
52 Ibid.
53 Robert Friedheim, 'Introduction: The IWC as a contested regime', in Robert L Friedheim (ed.), *Toward a sustainable whaling regime* (UWP, 2001) 3.
54 Andrew Miller and Dolšak Nives, 'Issue linkages in international environmental policy: The International Whaling Commission and Japanese development aid', (2007), *GEP* 1, 69.
55 Hoey, 'The Battle', 443.
56 Convention on the Law of the Sea, 10 December 1982, 1833 U.N.T.S. 397.
57 Conventional wisdom held that including the politically charged whaling issue in the *UNCLOS* negotiations could threaten the conclusion of the broader legal order that was being constructed. See Jefferies, *Marine Mammal Conservation*, 177.
58 Article 65 achieves this by allowing stricter regulation for marine mammals within the EEZ, which permits derogation from the guiding principle of optimum utilization of living marine resource management within optimum utilization.
59 Patricia Forkan, *The legislative history and interpretation of Article 65 of the Law of the Sea Convention*, Testimony Submitted 14 October 2003, < hsi.org/assets/pdfs/HSUS_testimony_LOS.pdf >, 4, citing to a Letter from the Honourable Elliot L. Richardson, Ambassador at Large, and Special Representative of the President of the Law of the Sea Conference, U.S Mission to the United Nations, to Patricia Forkan, Humane Society International (29 April 1980).
60 Ibid., 5.
61 Ibid., 3.
62 Elisa Morgera, 'Whale Sanctuaries', 322.
63 Ibid.
64 Kolmaš, 'International Pressure', 202.
65 Ibid.
66 David Day, *The Whale War* (Routledge 1987) 128.
67 Later in this chapter we discuss Japan's retreat from this objection and, alternatively, the use of Article VIII Special Permit whaling.
68 International Convention for the Regulation of Whaling, Schedule, paragraph 10(e). The moratorium only affected whale species regulated by the IWC, being great whales (all baleen whales sperm whales, minke whales), but not smaller whale types (all other toothed whales) and dolphins.
69 IWC34, Rep Int Whal Comn 34, 1984, s 8
70 IWC36, Rep Int Whal Comn 36, 1985, p. 56, s 2.5.1.

71 IWC36, Rep Int Whal Comn 36, 1985, p. 57, s 15.

72 The first IWC Resolution on whale watching was passed in 1993, and a second passed the next year at IWC46 directing the Scientific Committee to develop whale watching guidelines. The revision and enhancement of these guidelines became an ongoing agenda item in the following years (IWC, *Five Year Strategic Plan for Whale Watching 2011–2016* Nov 2011 at page 3). The IWC's first whale watching Strategic Plan was released in 2011, and it "set out a series of actions to support responsible and sustainable whale watching, including research and data collection, information sharing and capacity building" (IWC, Whale Watching: Providing an International Focus for the Whale Watching Sector, online <https://iwc.int/management-and-conservation/whalewatching>). Updated plans were released in 2016, 2018, and 2022, but utilizing and implementing the advice in these plans is not mandatory. Instead, "involvement in the actions of the Strategic Plan is on a voluntary basis and is intended to complement other national or international legislation, frameworks or plans that support the conservation and responsible non-lethal use of cetaceans" (IWC, *Five Year Strategic Plan for Whale Watching 2011–2016* Nov 2011, p. 7). Today, the main focus of the IWC's Whale Watching sub-committee is the continued development of an active and evolving digital Whale Watching Handbook, meant to "help the industry develop in a way that is sustainable in the long-term, for both the whale populations that are observed and the economies that depend on their presence" (IWC, *Whale Watching Handbook, A Comprehensive, Online Tool for Regulators, Industry and the General Public*, <https://iwc.int/whale-watching-handbook>). The Handbook supports policy makers at the national and regional level, whale watching businesses, and also members of the public seeking information on sustainable whale watching.

73 IWC36, Rep Int Whal Comn 36, 1985, p. 10, s 6.

74 Ibid., p. 67, s 90.

75 Ibid., p. 21, s 4.

76 Holm, 'Japan's Whaling Policy', 8. For further discussion of the Packwood-Magnuson Amendment of 1976, 16 U.S.C.§1821(1982), see: Susan Ghea, 'International Regulation of Whaling: The United States' Compromise', (1987), 27 Nat Resources J 931;

77 International Convention for the Regulation of Whaling, art. VIII.

78 Holm, 'Japan's Whaling Policy', 8. For further discussion, see Michael Strausz, 'Executives, legislatures, and whales: The birth of Japan's scientific whaling regime', (2014), 14 International Relations of the Asia-Pacific 455.

79 Charlotte Epstein, *The power of words in international relations: Birth of an anti- whaling discourse. Politics, science, and the environment* (MIT Press 2008) 82–83.

80 Holm, 'Japan's Whaling Policy', 9. For further discussion of political motivations, see: Kagawa-Fox, Midori, 'Japan's whaling triangle – The power behind the whaling policy', (2009), 29 *Japanese Studies* 401, 402-404.

81 IWC39 Rep Int Whal Com 38, 1988, p. 10, s 6.

82 Ibid.

83 Ibid.

84 Ibid.

85 IWC40, Rep Int Whal Comn 40, 1989, at p. 12, s 6.1

86 IWC44, Rep Int Whal Comm 43, 1993, p. 41.

87 Elisa Morgera, 'Whale Sanctuaries', 325.

88 William T Burke, 'Memorandum of opinion on the legality of the designation of the southern ocean sanctuary by the IWC', (1996), ODIL 315, 317. This memorandum was prepared for the Japanese Whaling Association. It was submitted to the International Whaling Commission at IWC48 in 1996. The argument was further developed by Burke in a subsequent article: 'Legal aspects of the IWC decision on the southern ocean sanctuary', (1997), 28 *ODIL* 313.

89 Elisa Morgera, 'Whale Sanctuaries', 328.

90 IWC44, Rep Int Whal Comn 43, 1993, p. 36, s 35.
91 Ibid., p. 37.
92 IWC45, Rep Int Whal Comn 44, 1994, p. 24, s 20.
93 Ibid., s. 20.1.
94 Ibid.
95 IWC46, Rep Int Whal Comn 45, 1995, p. 33, s 18.
96 Ibid.
97 IWC45, Rep Int Whal Comn 44, 1994, at p. 26.
98 Ibid.; IWC Resolution 1994–13, 'Resolution on research on the environment and whale stocks'.
99 International Whaling Commission, 'Environmental concerns', < https://iwc.int/management-and-conservation/environment > accessed 12 May 2023.
100 Ibid.
101 IWC48, Rep Int Whal Comn 47, 1997 Appendix 2 *Resolution on Whale Watching,* p. 48.
102 IWC Resolution 1998–5, 'Resolution on Environmental Changes and Cetaceans'.
103 IWC, "Bycatch", online < https://iwc.int/management-and-conservation/bycatch > accessed 12 May 2023
104 IWC Resolution, 1997–4, 'Bycatch Reporting and Bycatch Reduction'; IWC 49, Chairman's Report of the Forty-Ninth Annual Meeting, Rep Int Whal Comn 48, 1998. Bycatch has remained an important IWC agenda item. Most notably, 2016, at IWC66, the IWC unanimously endorsed a new Bycatch Mitigation Initiative (BMI). The BMI focuses on gillnet bycatch in small-scale fisheries, although "some work continues on bycatch in other gears and fisheries" (IWC, 'Bycatch'). The BMI 10-year strategic plan acknowledges that solutions are needed "on multiple levels including at the vessel, fleet, community, and national and international levels", and its primary aim is to raise awareness at the national and international level and to promote effective tools to mitigate and prevent bycatch deaths (IWC, 'Bycatch Mitigation Initiative Strategic Plan 2018–2023', (2018) IWC/67/CC/01 CC). The BMI's ultimate objective is to become an "active advisory body, providing information and knowledge transfer through awareness raising programmes, technical advice, capacity development/ training and input through regional and international fora" (*ibid.*, 4.). Moreover, the BMI presents technical advice on achievable bycatch mitigation "to countries (including those with endangered cetacean populations) upon request, with potential opportunities to provide advice through existing IWC work programmes such as Conservation Management Programmes and the Small Cetacean Task Team" (*ibid.*). The BMI is committed to working with the IWC's "Global Whale Entanglement Response Network", which was launched in 2011. While the BMI is focused on large scale knowledge mobilization outcomes, the Whale Entanglement program, launched in a partnership with the Center for Coastal Studies, is specifically focused on training responders to safely disentangle trapped cetaceans. Since its first training workshop in 2012, it has been delivered on five continents, and reached over 1,300 scientists, conservationists, and government representatives from more than 36 countries (IWC, 'Whale Entanglement: Building a Global Response', online < https://iwc.int/entanglement > accessed 12 May 2023).
105 IWC55, Rep Int Whal Comn 54, 2003, p. 10.
106 Gerry Nagtzaam, 'The temptation of the impossible normative contestation and deadlock at the International Whaling Commission', (2013), 8 *Appalachian Natural Resources LJ* 109, 127.
107 IWC, 'St Kitts and Nevis Declaration', Resolution 2006–1.
108 Ibid., 133.
109 Ibid., 135.
110 Ibid., 143

111 Matt DiCenso, 'Trouble on the high seas: A need for change in the wake of Australia v. Japan', (2016), 39 BC *International & Comparative Law Review* 13, 16.
112 Phillip J Clapham, 'Japan's whaling following the international court of justice ruling: Brave new world – Or Business as usual? (2014), 51 *Marine Policy* 238, 239.
113 Holm, 'Japan's Whaling Policy', 9.
114 Ibid., 8–9. Further, "Morishita argued that if Japan were to submit on whaling, other international fisheries arrangements could also be affected".
115 Judgment, ICJ GL No 148, ICGJ 471 (ICJ 2014).
116 Ibid., paras. 223–227.
117 Clapham, 'Brave New World', 239.
118 Ibid., para. 58.
119 Press Release, 'International court of justice whaling in the Antarctic (Australia v. Japan: New Zealand intervening) [Statement by Chief Cabinet Secretary, the Government of Japan]' (31 March 2014), < https://www.mofa.go.jp/press/danwa/press2e_000002. html >.
120 IWC, Report of the Scientific Committee, (2016), *Journal of Cetacean Research and Management* 17 (Suppl.), 72.
121 Virgina Morell, 'Japan says it will hunt whales despite science panel's opposition', (16 April 2015), *Science*, <https://www.science.org/content/article/japan-says-it-will-hunt-whales-despite-science-panel-s-opposition> accessed 14 September 2023.
122 Anastasia Telesetsky and Seokwoo Lee, 'After whaling in the Antarctic: Amending Article VIII to fix a broken treaty regime', (2015), 30 *Marine and Costal Law* 700, 721.
123 IWC66, 13.5.1 Report of the IWC Workshop to Develop Guidance for the Handling of Cetacean Stranding Events, p. 27
124 International Whaling Commission, 'Strandings initiative', <https://iwc.int/management-and-conservation/strandings/strandings-initiative> accessed 15 May 2023.
125 IWC, Strandings Initiative Draft 4-year Workplan 2021–2024, p. 12, s. 5.6.
126 IWC, 'Japan's opening statement to the 67th meeting of the International Whaling Commission', <https://archive.iwc.int/pages/download.php ref=7132&size=&ext=pd f&k=&alternative=-1&usage=-1&usagecomment=&usg=AOvVaw24imtP0VYOAm XnMm7J1rmt > accessed 15 May 2023.
127 Government of Japan, 'The way forward of the IWC: IWC reform proposal including a draft resolution and proposed schedule amendment', (2018), IWC/67/08.
128 Ministry of Foreign Affairs of Japan, Statement by Chief Cabinet Secretary (26 December 2018), <https://www.mofa.go.jp/ecm/fsh/page4e_000969.html> accessed 15 May 2023.
129 Chie Kojima, 'Japan's decision to withdraw from the international convention for the regulation of whaling', (2019), 4 *APJOCP* 93, 95–96.
130 3 March 1973, 993 U.N.T.S. 243.
131 Robert Steenkamp and Cameron Jefferies, 'In pursuit of the white whale of cooperation the ability of UNCLOS to steer the trajectory of (Future) commercial Japanese whaling operations', (2020), 5 *APJOLP* 245, 256.
132 CITES Secretariat, 'Introduction from the sea of Sei Whales (Balaenoptera Borealis) by Japan', (2018), Sc70 Doc.27.3.4.
133 For further discussion of this, see: Chris Wold, 'Japan's resumption of commercial whaling and its duty to cooperate with the International Whaling Commission', (2020), 35 *JELL* 87; Steenkamp and Jefferies, 'In Pursuit of the White Whale'.
134 IWC63, Rep Int Whal Comm 62, 2011, p. 7.
135 Steenkamp and Jefferies, 'In Pursuit of the White Whale', 259.
136 International Convention for the Regulation of Whaling, last preambular paragraph.
137 Judgment, para. 45.

138 See: Akiho Shibata, 'ICRW as an evolving instrument: Potential broader implications of the whaling judgment', (2015), 58 *Japanese Yearbook of International Law* 298 (2015).
139 IWC Resolution 2016–3, 'Resolution on Cetaceans and Their Contributions to Ecosystem Functioning'.
140 Ibid.
141 IWC, 'IWC workshop: The value of whales to the ecosystem' (2017), online: < https://iwc.int/resources/media-resources/news/IWCworkshop-the-value-of-whales-to-the-ecosystem >.
142 IWC Resolution 2018–5, 'The Florianópolis declaration on the role of the International Whaling Commission in the conservation and management of whales in the 21st century'.
143 Natalie J Lockwood, 'International vote buying', (2013), 54 *HILJ* 1 97, 99.
144 Alexander Gillespie, 'Good governance, corruption & (and) vote buying in international forums', (2004), 1 *NZYIL* 103.; IWC Resolution 2001–1, 'Resolution on transparency within the International Whaling Commission'.
145 Nagtzaam, 'The temptation of the impossible', 140.
146 Christian Dippel, 'Foreign aid and voting in international organizations: Evidence from the IWC', (2015), 132 *JPE* 1, 3.
147 Ibid.
148 Ibid., 2.
149 Ibid., 3.
150 'Japan walks out of IWC talks to avert vote on whaling sanctuary', *Thai News Service* (18 July 2011), originally posted in *Kyodo News International, Inc* (15 July 2011).
151 IWC, "Whale sanctuaries", < https://iwc.int/management-and-conservation/sanctuaries >, accessed 14 May 2023.
152 Sara Wissmann and Maurus Wollensak, 'Sometimes goodbyes are not forever: Japan's hypothetical re-accession to the international convention for the regulation of whaling', (2020), 34 *OYB* 164, 186–187.
153 Japan's Opening Statement to the 68th Meeting of the International Whaling Commission, IWC/68/OS/GO/Japan (2022), <https://archive.iwc.int/pages/view.php?ref=19788&k=e099d6a481> accessed 15 May 2023.
154 IWC/68/8.3/01/EN, Plenary Agenda Item 8.3, <https://archive.iwc.int/pages/download.php?direct=1&noattach=true&ref=19636&ext=pdf&k=> accessed 15 May 2023.
155 Ibid.
156 Chris Butler-Stroud, 'Antigua acknowledges aid for support of whaling', *Whale and Dolphin Conservation* (15 June 2015), <https://uk.whales.org/2015/06/15/antigua-acknowledges-aid-for-support-of-whaling/> accessed 15 May 2023.
157 Dillon De Shong, 'Antigua and Barbuda says it has not agreed to lift whaling ban', *LoopNews* (20 October 2022), online: <https://caribbean.loopnews.com/content/antigua-barbuda-says-it-has-not-agreed-lift-whaling-ban> accessed 15 May 2023.
158 IWC, 'Draft resolution on marine plastic pollution', IWC/68/8.1/01/REV2/EN, Plenary Agenda Item 8.1.
159 United Nations Environment Assembly of the United Nations Environment Programme, 'End plastic pollution: Towards an international legally binding instrument' (10 May 2022), UNEP/PP/OEWG/1/INF/1.
160 IWC, 'Draft resolution on marine plastic pollution'.
161 Anne Peters, 'The international convention for the regulation of whaling: Dead or alive?' (2021), 45 *AIL* 134.

5 'Opening up a Procedure'

Might the Re-adherence of Iceland to the International Convention for the Regulation of Whaling in 2002 Provide an Example for Japan to Follow?

Ed Couzens

Introduction

Over the lifetime of the 'ICRW', 1946 many states have adhered, then left and some have re-adhered. The first decade of the treaty's operation saw a high degree of unanimity, but as early as 1959, after the 11th Meeting of the IWC ('IWC11'), states like Japan, the Netherlands and Norway gave notices of withdrawal. The issue at the time was that states operating in the Antarctic had failed to agree on catch allocations, the three wanting higher allocations than they were given; both the Netherlands and Norway allowed their notices to take effect – Japan did not.[1] Norway rejoined in 1960, and the Netherlands in 1962 (but gave notice of withdrawal again in 1969, before re-adhering) – New Zealand, taking a new 'preservationist' approach, gave notice of withdrawal in 1968. Other states like Jamaica and the Seychelles have adhered, played (sometimes even important[2]) roles and then withdrawn. Iceland withdrew in 1992 and then re-adhered in 2002.

In 2007, at IWC59, Japan hinted strongly that it might withdraw – Japan's then Alternate Commissioner Joji Morishita suggesting that without seeing 'some tangible progress we will have to think of alternatives'.[3] At IWC67 in 2018, Japan put forward a proposal 'for the way forward of the IWC' – after this proposal was not adopted (the vote saw 27 in favour, 41 against and 2 abstentions), Japan stated that '[t]he result of the vote, [...] we all know, can be seen as equivalent to the denial of the possibility for the Contracting Governments with different views to co-exist with mutual understanding and respect within the IWC'.[4] Japan then indicated that:

> [...] if scientific evidence and diversity are not respected, if commercial whaling based on science is completely denied, and if there is no possibility for the different positions and views to co-exist with mutual understanding and respect, then Japan will be pressed to undertake a fundamental reassessment of its position as a member of the IWC, where every possible options will be scrutinized.[5]

Japan eventually did, at the end of 2018, give notice to withdraw and this notice took effect at the end of June 2019.

DOI: 10.4324/9781003250814-6

However, Japan saw value in the ICRW for decades and put into it both much effort and considerable expenditure – the door is not sealed against Japan's re-adhering. Indeed, in its statement following the vote on its proposal at IWC67 Japan indicated that

> [w]e continuously believe in the potential of the IWC as a forum of global governance for the conservation and management of whale resources, and therefore wish to continue its cooperation with the IWC in various ways to uphold the objectives enshrined in the Convention.[6]

Such re-adherence would probably, however, be controversial – it seems unlikely that Japan would re-adhere without compromises on both sides[7] needing to be made. One possibility is that Japan could seek to re-adhere with a reservation[8] to the moratorium on commercial whaling that is currently in place. If this were to be proposed, and especially if objected to by other Contracting Governments, then the experience of Iceland in 2001, 2002 and afterwards would be relevant.

The ICRW and Reservations

According to the Vienna Convention on the Law of Treaties, 1969 a state may, when signing, ratifying, accepting, approving or acceding to a treaty, formulate a reservation, unless (a) the reservation is prohibited by the treaty; (b) the treaty provides that only specified reservations, which do not include the reservation in question, may be made; or, (c), in cases not falling under sub-paragraphs (a) and (b), the reservation is incompatible with the object and purpose of the treaty.[9]

The ICRW makes it clear in Article V that objections (i.e.: reservations) may be made to amendments made to the Schedule. Per Article V.3, amendments 'shall become effective with respect to the Contracting Governments ninety days following notification of the amendment by the Commission to each of the Contracting Governments, except that':

> (a) if any Government presents to the Commission objection to any amendment prior to the expiration of this ninety-day period, the amendment shall not become effective with respect to any of the Governments for an additional ninety days; (b) thereupon, any other Contracting Government may present objection to the amendment at any time prior to the expiration of the additional ninety-day period, or before the expiration of thirty days from the date of receipt of the last objection received during such additional ninety-day period, whichever date shall be the later; and (c) thereafter, the amendment shall become effective with respect to all Contracting Governments which have not presented objection but shall not become effective with respect to any Government which has so objected until such date as the objection is withdrawn.

The ICRW created a management body, the IWC, which met annually from 1949 to 2012; and then biannually from 2014, with an unintended (Covid-19 pandemic-related) gap of four years between the 67th Meeting in 2018 and the 68th Meeting in 2022. The IWC may be attended by each Contracting Government represented by a Commissioner (and possibly also by Alternate Commissioners). There is also a Scientific Committee, which is composed of experts from around the world who do not formally represent Contracting Governments.

Background

In 1982, at IWC34, the IWC adopted a moratorium (technically, a 'zero quota'; but generally called 'the moratorium') on commercial whaling, to take effect from the 1985/6 whaling season. The moratorium was achieved through an amendment of the Schedule (amendments requiring a 75% vote in favour) to introduce paragraph 10(e):

Classification of Stocks

10. All stocks of whales shall be classified in one of three categories according to the advice of the Scientific Committee as follows:

… (e) Notwithstanding the other provisions of paragraph 10, catch limits for the killing for commercial purposes of whales from all stocks for the 1986 coastal and the 1985/86 pelagic seasons and thereafter shall be zero. This provision will be kept under review, based upon the best scientific advice, and by 1990 at the latest the Commission will undertake a comprehensive assessment of the effects of this decision on whale stocks and consider modification of this provision and the establishment of other catch limits.

Four Contracting Governments – Japan, Norway, Peru and the Union of Soviet Socialist Republics (its place now taken by the Russian Federation) – lodged objections (i.e.: entered reservations) to paragraph 10(e) within the prescribed period. The paragraph came into force on 3 February 1983 for all other Contracting Governments. Subsequently, however, there have been two withdrawals of these reservations. Neither Norway nor the Russian Federation have withdrawn their reservations, and the paragraph is not binding upon these two Governments. Two Contracting Governments, however, did withdraw their objections – Japan and Peru.[10]

Japan's withdrawal was more complicated than Peru's blanket withdrawal. Japan withdrew its objection in respect of commercial pelagic whaling with effect from 1 May 1987; withdrew its objection in respect of commercial coastal whaling for minke and Bryde's whales with effect from 1 October 1987; and withdrew its objection in respect of commercial coastal sperm whaling with effect from 1 April 1988. Japan's reasons for withdrawal of its reservation are interesting, concerning claims that the withdrawal was made due to threats of the loss of fishing quotas in

US waters, but are not directly relevant to the present subject matter (except insofar as they go to general issues of lack of trust between different Contracting Governments) and are not discussed here.

Iceland did not lodge an objection. This is perhaps surprising because Iceland did, at IWC34, make its disapproval of the moratorium clear; stating, in its oral response to the adoption of the moratorium, that:

> [a]doption of a moratorium on all commercial whaling based on strength of votes but not on scientific findings nor recommendations of the Scientific Committee, nor with any regard to the economical or social importance of whaling to the whaling nations, is in our opinion most certainly contrary to the fundamental objectives and purposes of the Commission.[11]

Iceland's Withdrawal

The Government of Iceland gave formal notice in December 1991 that it would withdraw from the ICRW; such withdrawal to take effect on 30 June 1992. According to Iceland, the decision to withdraw from the IWC was made because '[i]n recent years the Commission has failed to reach solutions acceptable to Iceland'; and because Iceland 'traced these failures to structural deficiencies in the IWC, disrespect for the rules under which it should work and disregard for the advice of its Scientific Committee' – according to Iceland, it 'thus concluded that the Commission was fundamentally flawed'.[12] Iceland then stated that it (i.e.: Iceland) 'has a general concern for the environment'; that it is 'overwhelmingly dependent on the exploitation of the living resources of the ocean'; and that it was 'encouraged by the Rio endorsement of the principle of sustainable utilisation of marine living resources'.[13]

The Minister of Fisheries for Iceland stated that:

> The economic and social fabric of this island nation is overwhelmingly dependent on the health and productivity of the surrounding marine environment. Whales have an important ecological role in the Icelandic Exclusive Economic Zone; they consume more than the amount of seafood that our fishermen harvest. Whales must, therefore, be treated in the same manner as other resources, subject to the same management principles. ... It should not be difficult to understand why this Government must respond to the grim reality that the International Whaling Commission is no longer a viable forum for international cooperation on the conservation and management of the whale populations in our region. It is clear that Iceland has no choice but to seek cooperation in this field through the establishment of a new organization for the North Atlantic.[14]

'In sum', the Minister of Fisheries said, 'the Icelandic Government [has] concluded that the IWC is, and will remain, an anachronistic and ineffective organization'.[15] What withdrawal meant was that Iceland was left free to operate on its own or to work with other states in different fora. Some envisaging of different organisations can be found in the United Nations (UN) Convention on the Law of the Sea, 1982, which, in its Article 65, uses the plural sense of 'organisations' through which Parties are to work in the conservation and management of cetaceans:

Article 65 Marine mammals

Nothing in this Part restricts the right of a coastal State or the competence of an international organization, as appropriate, to prohibit, limit or regulate the exploitation of marine mammals more strictly than provided for in this Part. States shall cooperate with a view to the conservation of marine mammals and in the case of cetaceans shall in particular work through the appropriate international organizations for their conservation, management and study.

Iceland, then, in Davies' words, 'took the dramatic step to disengage completely from the Whaling Convention and instead to cooperate with other like-minded parties within NAMMCO'.[15] Iceland, per Davies, became a founding member of the North Atlantic Marine Mammal Commission (NAMMCO), which was established by the 1992 Agreement on Cooperation on Research, Conservation and Management of Marine Mammals in the North Atlantic.[16] Adopted by Ministers responsible for fisheries management in the Faroe Islands, Greenland, Iceland and Norway, NAMMCO's goal is to 'contribute through regional consultation and cooperation to the conservation, rational management, and study of marine mammals in the North Atlantic' and especially to 'endorse the sustainable utilization of cetaceans in the North Atlantic'.[17] However, Davies' judgment is that 'such withdrawal diminished Iceland's international voice and, in doing so, reduced the potential impact of the pro-whaling lobby at the international level'; and, 'as a consequence, in June 2001, Iceland attempted to rejoin the IWC'.[18]

Iceland's First Attempt to Re-adhere

At the IWC's 53rd Meeting ('IWC53'), in London in 2001, Iceland sought to rejoin the IWC (more accurately, to re-adhere to the Convention which would mean its once more becoming a Contracting Government to the IWC). Iceland explained that it had withdrawn in 1992 because it had believed that the IWC 'was no longer operating in accordance with the Convention and had become a non-whaling commission rather than a whaling commission'.[19] Iceland explained, however, that it now felt that there were signs within the IWC of support increasing for sustainable whaling; that it had, therefore, rejoined in order to 'have an influence on the discussions taking place'; and that it had since leaving been 'urged' by 'a number of countries, both for and against commercial whaling', to rejoin.[20] However, what caused controversy is that Iceland sought to rejoin with its instrument of adherence being made expressly conditional on a reservation to the commercial whaling moratorium, as found in paragraph 10(e) of the Schedule to the ICRW. Iceland claimed that it had the right under international law to adhere with a reservation and that there was 'no legal basis for rejecting' its re-adherence; and explaining that Iceland considered it 'outside the competence of [the] IWC to take a decision on Iceland's reservation by voting on it'.[21]

In its written Opening Statement to IWC53, Iceland had argued that:

> The Convention itself grants Contracting Governments the right not to be bound by this paragraph [10(e) of the Schedule] by way of objecting to it as Iceland has done with its reservation. How can something, which is allowed in the Convention be incompatible with the object and purpose of that very same Convention? The answer is simple: it can not.[22]

Iceland then argued that 'international relations are based on the rule of law' and that:

> Iceland can not believe that any country present at this 53rd annual meeting of the IWC will actually sidestep this principle in favour of political objectives. If this happens, States are allowing politics to override the rule of law. This would not only undermine the credibility of the IWC but also that of international relations in general.[23]

In its written Opening Statement to IWC53, Antigua and Barbuda argued that:

> ... to have a vote by the IWC on the reservation made by Iceland is an absurdity, which this organisation would not easily live down. It would confirm the suspicions in the minds of many: that the IWC is an organisation that tramples on the sovereign rights of member countries in rampant pursuit of the extremist agenda of special interests.[24]

The Solomon Islands, in its written Opening Statement, said that it had:

> ... watched over the years with great disbelief, the path troddened by the IWC, ignoring its purpose and mandate, which is to ensure conservation of whale species and the orderly development of whaling, but instead is being steered towards the total conservation of whales, despite the fact that certain stocks are well over their sustainable levels.[25]

Opinion amongst other Contracting Governments was divided. Australia opposed Iceland's rejoining with a reservation; arguing that it was the IWC which would have to decide whether this was acceptable or not. The United States supported this, pointing out that Iceland had had the opportunity to lodge a reservation in 1982, while a member, but had not done so; and that 'acceptance of this reservation now would undermine the commercial whaling moratorium'. Argentina, Finland, Germany, Ireland, Italy, Monaco, The Netherlands, New Zealand (which argued that the Convention does not provide for reservations; and that 'the reservation to the Convention requires the acceptance of the competent organ, and that therefore the Commission can and must decide'), Spain, Sweden and the United Kingdom expressed views similar to those of Australia.[26]

Japan argued against the Australian position; suggesting that 'acceptance or not of the reservation is a decision for each Contracting Government'. Antigua and Barbuda, Grenada, the Republic of Guinea, Iceland, Norway, the Russian Federation and St Lucia spoke in support of Iceland; arguing that Iceland 'was within its rights' to adhere with a reservation, and that the Commission 'did not have competency' to refuse it.[27]

Some Contracting Governments were either undecided or deferred expressing a view. Switzerland argued that the IWC is not an international organisation 'as it does not qualify as an autonomous body that possesses an independent legal personality'; and that a vote on the present issue would itself be against international law. France explained that it would object to Iceland's reservation; but that it did not believe that the Commission was entitled to vote on the issue. Denmark considered the issue to be so complicated that it should not be decided at that stage; and that a vote should not be taken.[28]

The Chair[29] then ruled, in the absence of any clear view, that the Commission did indeed have the competence to determine the status of Iceland's proposed adherence with a reservation. This ruling was to become known as 'the London rule'. The vote (on whether the Chair's ruling should be upheld or not) saw 19 votes in favour, 18 against and 1 abstention (Austria). Iceland indicated that it 'regretted' the vote and explained that it considered it to be an 'illegal vote'.[30]

The Chair then proceeded to the substantive motion (which was an Australian/ United States motion that Iceland's adherence with a reservation not be accepted). However, all of Antigua and Barbuda, China, Dominica, Grenada, Iceland, Japan, the Republic of Guinea, Morocco, Norway, Panama, the Solomon Islands, St Kitts and Nevis, St Lucia and St Vincent and the Grenadines indicated that they would not take part, as they considered the vote to be 'illegal'. The motion, put to the vote, received 19 votes in favour and 0 against; with 3 abstentions and 16 countries not participating.[31]

The Chair then ruled that Iceland would, thenceforth, be 'invited to assist as an observer'. This was contentious too, with Japan opposing the ruling. Put to the vote, it was carried with 18 for, 16 against and 3 abstentions. Iceland, however, 'indicated its intention to continue to participate in the meeting as a Contracting Government'.[32]

Iceland's Second Attempt to Re-adhere

At IWC54, held in Shimonoseki in May 2002, Iceland again deposited an instrument of adherence with a reservation. If the IWC were to fail to accept the reservation, since Iceland made it an integral part of its instrument of adherence, Iceland would not be considered a contracting party to the IWC Convention.

Deposit was made with the USA – the depository government. The USA circulated at the Meeting a statement expressing its view as being that:

> The United States considers the Icelandic reservation to constitute a proposed amendment to the Schedule. In accordance with the terms of the IWC

Convention, we view that reservation as having no legal effect until it has been accepted by a three-fourths majority of those members of the International Whaling Commission (IWC) voting.[33]

As will be seen, however, this view on the need for a 75% vote was superseded by events.

Iceland stated that it now regarded itself as a member of the IWC, despite the events and the Chair's ruling of the previous year; alternatively, that if it was considered that its reservation (and thereby its adherence) had been rejected at IWC53 (and were therefore not in effect), then its new instrument of adherence of 2002 'must be regarded as a fully valid new instrument of adherence' – and that, having deposited this instrument, it was a 'new Contracting Government until it is challenged'.[34]

Antigua and Barbuda noted that it considered that the Commission has no authority to deny Antigua and Barbuda from accepting Iceland as a new member; 'or to interfere with treaty relations between States'.[35] Denmark, after being undecided the previous year, gave its view that the IWC is not competent to decide on this issue; which Denmark felt should be a 'bilateral matter between Iceland and those individual Contracting Governments having problems with Iceland's reservation'. China, Grenada, the Republic of Guinea, Japan, Norway, the Republic of Palau, the Russian Federation, St Kitts and Nevis and St Lucia stated that they recognised Iceland as a member.[36]

The United States, however, gave its view that Iceland 'wanted to be the sole judge of whether to exercise its reservation in the future' and argued that 'if Iceland does not like the commercial whaling moratorium', then 'it should join the IWC without reservation and work towards having the moratorium lifted'. Australia, Germany, Ireland, Italy, Mexico, New Zealand, Spain and the UK argued along similar lines.[37]

Antigua and Barbuda argued, however, that 'the treaty relations which exist among IWC members are not between the individual Contracting Governments and the Commission but between the individual Contracting Governments themselves'; and that each Party had the right to object, in which case the reservation would have no effect as between those Parties, or the right to accept the reservation. Antigua and Barbuda argued further that there was 'adequate precedent regarding reservations' and that, in the past, they had been addressed by individual Contracting Governments instead of by the Commission.[38] On the issue of previous reservations, Japan pointed out that Argentina and Ecuador had, at their adherences, made reservations relating to territorial waters and noted that the IWC had not, then, intervened with respect to these reservations as it was now doing with Iceland.[39]

The Chair then repeated his ruling of 2001 and invited Iceland to participate as an Observer; at which point, Japan and Norway challenged the ruling, and also the competence and authority of the Chair to make such a ruling. A vote on the Chair's ruling was then held; but the challenge was not successful, with 20 votes for and 25 against, and the Chair's ruling was upheld. On the Chair's competence then being put to the vote, there were 17 votes in support of the challenge, 24 against and 3 abstentions.[40]

Iceland then withdrew; after making a statement (a 'formal declaration') explaining that it 'considered all attempts not to recognise it as a member' to be 'illegal, therefore not affecting its status as a member'. According to Iceland, there had been breaches of the general principles of international law, of the ICRW and of the IWC's Rules of Procedure; it argued specifically also that the United States (as depository government) had misused its position in 'not notifying Iceland as a member' of the ICRW, that the Chair had acted contrary to the ICRW, and that a majority of members had 'violated general principles of international law' and the ICRW.[41]

Context

At this point, it is essential to understand a different issue, the events surrounding Aboriginal Subsistence Whaling at IWC54 in May 2002. 2002 was a year in which the Russian Federation and the United States were both seeking renewals of their five-year quotas for the taking of bowhead and gray whales in the Bering and Chukotka Seas by their Aboriginal (Indigenous) peoples. As amending or renewing the quotas meant amending the Schedule, a 75% vote was required. In an unexpected move at the 54th Meeting, Japan led resistance to renewal of the quotas based on uncertainty over stock numbers. The quotas were not renewed.[42]

What happened at IWC54 was that, unexpectedly, Japan noted that it considered the bowhead stock 'to be in a very dangerous situation' and eventually – despite a 'logrolling manoeuvre' proposal by Japan, which might have seen approval of the quota if politically linked to the approval of a quota for Japanese Small Type Coastal Whaling – that it 'could not support a five-year block quota over concern for the status of the bowhead stock'.[43]

Russia and the United States then called an intersessional meeting, the fifth Special Meeting of the IWC, which was held in October 2002.

Iceland's Third Attempt to Re-adhere

The Special intersessional Meeting was held at the request of the Russian Federation and the United States, in order to repeat their request for a Schedule amendment to allow for the granting of (for the USA) an aboriginal subsistence take of bowhead whales from the Bering-Chukchi-Beaufort Seas stock; and (for the Russian Federation) of gray whales in the seas off its remote Chukotka Region. Iceland took advantage of the extraordinary Meeting to make a third attempt to re-adhere.

Iceland deposited, on 10 October 2002, an instrument of adherence with the United States (as depositary government). Again, the reservation to the moratorium was an integral part of the instrument; however, Iceland had made a 'political' concession by committing not to authorise whaling 'for commercial purposes by Icelandic vessels before 2006' and not to 'authorise such whaling while progress is being made in negotiations on the [Revised Management Scheme, or 'RMS'].[44] Iceland noted further that '[u]nder no circumstances will whaling for commercial purposes be authorized without a sound scientific basis and an effective management and enforcement scheme'.[45]

The Chair decided that, as in previous years, it should first be decided whether or not the Commission has the competence to decide the issue; before, second, deciding whether or not to allow adherence with a reservation.[46]

Iceland's instrument of adherence to the ICRW, deposited on 10 October 2002, stated that:

> Iceland 'adheres to the aforesaid Convention and Protocol with a reservation with respect to paragraph 10(e) of the Schedule attached to the Convention'.[47]

What happened next is extremely confusing. After a number of points of order had been taken[48] on the basis that the matter had been resolved at IWC53 and upheld at IWC54; and that the instrument of adherence contained the same reservation, the Chair (Bo Fernholm, the Commissioner for Sweden) began by ruling that the procedure used in London should be followed; i.e.: further points of order[49] were considered, with the IWC's competence to decide the issue of Iceland's re-adherence with a reservation challenged.[50] The Chair ruled, in the absence of consensus on how to proceed, that the IWC would follow the procedure adopted at IWC53, in effect that the IWC has the competence to determine the legal status of Iceland's reservation. At IWC53, after a vote, the IWC did not accept Iceland's reservation; and Iceland was invited to participate as an observer.[51]

Brazil then challenged the Chair's ruling. Brazil argued that 'there was nothing new in the substance of the adherence' and that 'it considered that the issue of competency had already been decided'.[52] There were then four votes taken.

A Vote on a Challenge to the Chair's Ruling That the IWC53 Procedure Should Be Followed

A vote was held on the challenge to the Chair's ruling that the procedure adopted at IWC53 should be followed. The vote was successful and the ruling was overturned: 21 to 16.[53]

A Vote on a Challenge to Vote Immediately, Following the IWC53 and IWC54 Procedures

The United Kingdom then formally proposed that, in consequence of the overturning of the Chair's ruling, the procedure adopted at IWC54 should be followed. The Chair proposed that this immediately be put to a vote. Norway, however, challenged the Chair's ruling (of putting the United Kingdom's proposal to a vote immediately) on the ground that the IWC did not have the competence to make such a decision.[54]

Put to a vote, Norway's challenge to the Chair's ruling was defeated – the vote tallies being 18 each way.

*A Vote on a Challenge (by Mexico) to the Rule That Iceland Should Be
Allowed to Vote*

Following the second vote, the Chair ruled, as he had at IWC54 earlier in 2002, and
on the basis that Iceland's new instrument of adherence contained the same reservation (just with an additional declaration), that the position decided from before
continued to apply, i.e.: that the IWC has the competence to determine the legal
status of Iceland's reservation; that the IWC did not accept Iceland's reservation;
and that Iceland was invited to participate as an observer. The Chair indicated that
he felt bound by the previous decisions 'unless and until' the Commission were to
decide otherwise.[55]

Antigua and Barbuda challenged this ruling, arguing that what the IWC should
be voting on was not a procedure used at the last meeting but rather on the competence of the IWC to decide on Iceland's reservation. The Chair then indicated
that the competency issue had already been voted on and that the issue now was
his ruling that the procedure followed at IWC54 should again be followed. Various
views followed – including that of the United Kingdom that 'it was implicit in any
votes taken in respect to it, that Iceland should have no vote since the Chair's ruling
stands until it is defeated', and that the United Kingdom wished to be given clarifi-
cation on this. The Chair, after indicating that he agreed with the view of the United
Kingdom, then instructed that a vote should begin on Antigua and Barbuda's challenge to his ruling. He indicated that (in accordance with the IWC54 procedure)
Iceland would not be called on to vote.[56]

There was then debate on whether Iceland should be allowed to vote or not.
Views expressed included that of Antigua and Barbuda, which questioned why
Iceland should not be allowed to vote when the IWC54 decision had not yet
been accepted; and that of Sweden, which argued that until the IWC had 'made
a definitive decision' Iceland was entitled to participate in voting, because of
its new instrument of adherence. Antigua and Barbuda argued that there should
be a vote on whether to allow Iceland to participate in voting before there was
a vote on the challenge to the Chair's ruling. The Chair then indicated that he
'considered that it might be fair' to allow Iceland to participate in the vote, as
Iceland 'had been allowed to participate in the voting so far'. Australia argued,
however, that the Chair had already ruled that, in view of past decisions, Iceland was an observer, that this ruling was subject to challenge, that Iceland
should therefore not be entitled to participate in the vote on the challenge, and
that the vote should proceed immediately. Iceland responded by questioning
'whether Australia was challenging the Chair's ruling that Iceland would have
the right to vote' and stating that if that were the case then 'there would need to
be a vote on that'. Ireland then 'noted' that the Chair had ruled that the previous decision stood, and that 'since this ruling had been challenged it should go
to a vote'. The Chair then stated that, 'having listened to the different views
expressed', he 'ruled that Iceland should be allowed to participate in the vote'.
The Chair had, effectively, changed his mind on the issue. Mexico challenged
this ruling. Various views[57] were then put forward on the status of the challenge
and of Iceland.[58]

Mexico's challenge to the ruling by the Chair that Iceland be allowed to vote saw another 18–18 split – and therefore the challenge failed. Sweden abstained in the vote, explaining that it had done so because its view was that Iceland had the right to vote until the Commission had decided otherwise.

A Vote on a Challenge to the Rule That the Decisions on Iceland's Re-adherence (Made at IWC53 and IWC54) Be Upheld

The Chair then instructed that a vote be held on the challenge by Antigua and Barbuda (to his ruling that the decisions in London and Shimonoseki be up-held). In other words, Antigua and Barbuda challenged the Chair's ruling that, as decided at IWC53 and IWC54, first, the IWC has the competence to decide the legal status of Iceland's reservation; second, that the Commission does not accept Iceland's reservation; and, third, that Iceland is invited to assist as an ob-server. Iceland, supported by Antigua and Barbuda, noted that this vote 'involved the issue of competence' and 'supporting the Chair's ruling also supported the view that the Commission has the competence to decide the issue of Iceland's membership'.[59] The challenge to the Chair's ruling was then upheld, by a vote of 19 to 18.

Oddities included that Sweden did not abstain this time, but voted in favour of upholding the challenge to the (Swedish) Chair's ruling.[60] Iceland voted (unsur-prisingly, in favour of upholding the challenge).

Reactions

Something that might be noted is that the fifth Special Meeting was not well at-tended. In total, 48 Contracting Governments attended IWC54 in 2002 – but 12[61] of these did not attend the Special Meeting. It is, of course, impossible to know what results might have arisen had more attended. Much in the IWC depends on small voting margins.

In reaction, Ireland noted that 'the Commission had voted to accept Ice-land as a member with its reservation' – Ireland then 'welcomed Iceland as a member' but 'indicated that it would submit a formal objection on a bilateral basis'.[62]

On the issue of Iceland's vote, Brazil indicated that it 'considered it incorrect to allow Iceland to vote in a vote that was basically upholding previous decisions'; and that 'it considered that the outcome of the vote had been seriously undermined since Iceland was voting in its own interest'.

Norway, however, responded and advised that it 'considered this normal'. Mex-ico described Iceland's voting on the Chair's ruling as 'illegal' but stated also that it would abide by the decision – and that it 'wished to put on the record Mexico's objection to Iceland's reservation'. Australia explained that it objected, and would register its objection formally, to Iceland's reservation; and to Iceland being al-lowed to vote on the matter.[63] Contacting Governments such as Australia, France, Italy, Mexico, New Zealand, the Netherlands and the UK indicated that they would object formally to Iceland's reservation; some, such as Monaco and the United States, indicated that they would object bilaterally.

The United States 'noted that its difficulty was not with Iceland but with [Iceland's] reservation and the precedent' set thereby – however, said the United States, 'now that the Commission had taken this decision', the United States 'looked forward to working with Iceland in a constructive manner'.[64] New Zealand advised that it considered that the decision had 'opened up a procedure that would enable countries once bound by a treaty to leave the organisation then to return making reservations to whatever they find objectionable'. New Zealand explained that it was gravely concerned not only for the IWC's integrity but also for all multilateral environmental agreements. New Zealand's concern is worth remembering.

The 'integrity of all multilateral environmental agreements'

The issue has arisen before of a state Party to a treaty not entering a reservation when entitled to do so, then subsequently withdrawing from the treaty before attempting to re-adhere with the reservation it had not previously made. The high-point of this practice can probably be seen in the actions of Trinidad and Tobago in 1998, when the state denounced the First Optional Protocol to the International Covenant on Civil and Political Rights (ICCPR), 1966, and then immediately re-acceded with a reservation purporting to exclude death penalty cases – leading immediately to objections by seven states, and ultimately to a second denunciation by Trinidad and Tobago.[65]

In 1999, the UN Human Rights Committee (the UNHCR) established under the ICCPR made a decision on admissibility in *Rawle Kennedy v Trinidad and Tobago*.[66] Kennedy was a death row prisoner. Trinidad and Tobago had not lodged a reservation to the First Optional Protocol to the ICCPR which allowed death row prisoners a right of appeal beyond the state itself – in May 1998, Trinidad and Tobago denounced the First Optional Protocol and then immediately (on the same day!) re-acceded, including a reservation which precluded the right of appeal for death row prisoners.[67] The UNHCR decided that the reservation made by Trinidad and Tobago on re-accession was invalid – striking out the 'offending' reservation and then finding that Trinidad and Tobago remained bound as a state Party to the First Optional Protocol.[68] The UNCHR found that the reservation was contrary to the object and purpose of the First Optional Protocol and that it was accordingly invalid.[69]

Relevant to the discussion of Iceland's re-adherence, the Office of the High Commissioner of Human Rights (OHCHR) decided that 'it is for the Committee, as the treaty body to the [ICCPR] and its Optional Protocols, to interpret and determine the validity of reservations made to these treaties'.[70] Further, the OHCHR considered that reservations could be made to the First Optional Protocol, 'as long as they are compatible with the object and purpose of the treaty in question'; and that '[t]he issue at hand is therefore whether or not the reservation by the [s]tate [P]arty can be considered to be compatible with the object and purpose of the Optional Protocol'.[71]

The OHCHR then determined that:

The function of the first Optional Protocol is to allow claims in respect of [the Covenant's] rights to be tested before the Committee. Accordingly, a reservation to an obligation of a State to respect and ensure a right contained in the Covenant, made under the first Optional Protocol when it has not previously been made in respect of the same rights under the Covenant, does not affect the State's duty to comply with its substantive obligation. A reservation cannot be made to the Covenant through the vehicle of the Optional Protocol but such a reservation would operate to ensure that the State's compliance with that obligation may not be tested by the Committee under the first Optional Protocol. And because the object and purpose of the first Optional Protocol is to allow the rights obligatory for a State under the Covenant to be tested before the Committee, a reservation that seeks to preclude this would be contrary to object and purpose of the first Optional Protocol, even if not of the Covenant.[72]

The necessary implication of this finding was that, in effect, the reservation lodged by Trinidad and Tobago on re-accession was invalid and had to be struck out; that Trinidad and Tobago therefore became a Party to the First Optional Protocol without the reservation; and that the OHCHR was entitled to hear the communication from Mr Rawle Kennedy.

The case has certain implications for discussion of Iceland's re-adherence. On one hand, it might be objected that the cases can be distinguished. In this regard, Iceland made it clear that its re-adherence was conditional on acceptance of its reservation – so that the situation could not arise, as it had in the *Rawle Kennedy* case, of the state purporting to lodge a reservation being found still to be a Party to the treaty without the reservation.[73]

On the same hand, it might be pointed out that human-right–based treaties (which have been described as having a 'unique character'[74]) might not follow quite the same rules as do others. As Fournier puts it, '[h]uman rights treaties differ from other multilateral treaties, since they are not reciprocal and do not imply a synallagma of duties between the contracting parties'; and the duties that states Parties incur exist 'in fact not towards the other contracting [P]arties, but towards their own citizens'. [75]

On the other hand, as it was put by Norway in respect of Trinidad and Tobago's attempted re-accession with a reservation, Norway 'considers the denunciation of the [First] Optional Protocol followed by a re-accession upon which a reservation is entered, as a circumvention of the established rules of the law of treaties that prohibit the submission of reservations after ratification'.[76] As McGrory comments, 'Trinidad and Tobago's strategy of denunciation with simultaneous re-accession with reservation is suspect under [...] the Vienna Convention on the Law of Treaties'.[77] This strategy, per McGrory,

functionally ... achieves the same outcome as a partial denunciation of the treaty – an act seemingly prohibited by Article 44[78] of the Vienna Convention.[79]

This tallies with New Zealand's concern 'not only for the IWC's integrity; but for all multilateral environmental agreements'.

McGrory poses the general question '[h]ow […] can we retain treaty integrity while adhering to the fundamental consensual nature of state participation in treaties?'.[80] Gillespie, writing specifically about Iceland's re-adherence to the ICRW, suggests that while Iceland's re-adherence with a reservation was successful, with the implication that this can be seen as having undermined the IWC's integrity, 'a certain amount of integrity was retained by the Commission as it ultimately controlled the process and the overall decision was not dictated to the Commission on a bilateral basis'.[81]

It has been argued, by Aust, that for 'a party to withdraw and then re-accede solely for the purpose of making a reservation that it did not make originally and that if made as a late reservation is unlikely to have been accepted, *is open to the most serious objection*' (emphasis added).[82] While a late reservation would almost certainly not have been accepted, Aust's suggestion might be distinguished on the basis that Iceland arguably did not re-adhere to the ICRW for the sole purpose of entering a reservation – Iceland made the argument, on its first attempt to re-adhere in 2001, that it could discern a change of approach in the IWC. Iceland had not been a Contracting Government for almost a decade before first attempting to re-adhere.

The Legal Position on Reservations

Generally speaking, while reservations still have a place in international law, and will have a place for many years to come, they are being 'squeezed out' – more recent environmental treaties[83] regularly prohibit them. As Gillespie notes, after canvassing a number of treaties in the issue-areas of ozone, climate change, biodiversity, Antarctic protection, transboundary movement of hazardous waste, desertification, fisheries and general oceans governance, that '[i]t is possible to assert that the majority of important international law documents are moving toward a situation of not allowing reservations'.[84]

A fundamental question which has not been dealt with directly by the (few) commentators who have written on the implications of Iceland's re-adherence with a reservation is that of why a time limit is imposed at all – 90 days in the case of the ICRW.

This could be argued to be relevant when considering the Vienna Convention on the Law of Treaties, raised above, that a state may formulate a reservation unless the reservation is prohibited by the treaty; or the treaty provides that only specified reservations, which do not include the reservation in question, may be made; or that the reservation is incompatible with the object and purpose of the treaty – in summary. The most logical reason for the imposition of a time limit is to create certainty and to avoid Parties moving in and out of positions. If one then sees the time limit included in the text of the ICRW as being integral to the right to lodge a reservation,[85] then this would provide a reasonable barrier to allowing a Contracting Government to withdraw and then re-adhere with a reservation.

Objections and the Current Position

At the next Meeting, IWC55 in 2003, Italy, Mexico and New Zealand made statements in which they declared that they considered either that Iceland's reservation, in terms of paragraph 10(e) of the Schedule, was invalid, being 'incompatible with the object and purpose of the Convention'.[86] Italy and Mexico[87] both stated, in addition, that they 'do not recognise Iceland as a party to the Convention or as a member of IWC, nor its right to vote' – and called upon Iceland to withdraw the reservation. According to New Zealand, it 'does not accept the Convention as being in force between itself and Iceland'. Iceland responded that it had received 'diplomatic notes' from these three Contracting Governments, and from others, and that it 'considered the statements made at the meeting to not be relevant'.[88]

> As a footnote to the Schedule, it is currently noted that:
> [t]he Governments of Argentina, Australia, Brazil, Chile, Finland, France, Germany, Italy, Mexico, Monaco, the Netherlands, New Zealand, Peru, Portugal, San Marino, Spain, Sweden, UK and the USA have lodged objections to Iceland's reservation to paragraph 10(e).[89]

Although the contracting governments which objected to Iceland's rejoining as a member have not withdrawn those objections, neither have they at any recent meeting objected to Iceland's participating, speaking, voting and so forth. The opportunity arises early at every meeting under the standing agenda item of 'Credentials and voting rights', but Contracting Governments have not since objected to Iceland's voting. Iceland resumed commercial whaling in October 2006, its first kill being that of a fin whale[90] and has since taken whales without formal objections being made to its right to do so.[91]

Conclusion

Japan surprised many – like the present author – by withdrawing from the ICRW in mid-2019. Notice to withdraw was given just three months after IWC67 in Brazil, and at that meeting Japan attended with a large delegation; appeared as engaged as ever before; and two new Contracting Governments – Liberia and São Tomé et Príncipe – adhered, both making Opening Statements which indicated that they would be most likely to support Japan's and the Sustainable Use Group's views. Writing with no special insight into the politics of Japan's withdrawal, it is possible to look back and suggest that Japan's decision to leave was in fact made a long time before IWC67; but, the value of understanding historical events aside, such speculation seems pointless at this stage. More relevant is to consider whether Japan might re-adhere and whether any such re-adherence would be made conditional on a reservation being lodged to the moratorium on commercial whaling. The ICRW and the IWC have a long history and have made extraordinarily valuable contributions to understanding whales and whaling. Without Japan, a major role player, this contribution is likely to be diminished in the future. Hopefully, Contracting Governments on the 'anti-whaling side' would recognise this too and they would support re-adherence.

The intended value of the present contribution has been to consider whether Iceland's re-adherence in 2002 with a reservation to the moratorium might provide a model that could be adopted by Japan were it to re-adhere. The best answer is probably that it would, but that this would not be uncontentious.

Were such an attempt to re-adhere to happen, it is not clear how other Contracting Governments would react. Most, if not all, would probably welcome Japan back. Many, however, would be likely to object were such re-adherence to be made conditional on a reservation – especially as they would not wish to set a precedent that more states might rely upon in the future (a second such re-adherence would carry considerable weight as a precedent for a third). Iceland's successful re-adherence would be put forward as a model and would certainly be re-examined closely. As a close and relevant example, Japan and others would probably argue that Iceland's experience sets a precedent – Contracting Governments on the 'anti-whaling side' would probably disagree.

The example of the *Rawle Kennedy v Trinidad and Tobago* case was not raised in debate in the IWC in 2001 and 2002 but is relevant. That Trinidad and Tobago withdrew and then re-acceded *on the same day* is an important distinguishing factor. Crucial might be the timing of when Japan made such a re-adherence bid – 'too soon after leaving' would expose Japan to the accusation that it had left intending to return with a reservation; leaving 'too long a gap' before re-adhering carries the danger that for both 'sides' momentum might be lost, the *status quo* stabilize, re-adherence might never happen, and the IWC might lose relevance as a decision-making organisation.

Given the 'break' the IWC took from Meetings in the 'Covid-19 Pandemic years' – the next meeting after IWC67 in 2018 was IWC68 in 2022 – by the time of IWC69 (due in 2024) it will be difficult to make the case that Japan withdrew *solely in order to* follow Iceland's model and re-adhere with a reservation.

Gillespie concludes that the most important 'takeaway' from the events in the IWC of 2001 and 2002 may be that the decision – whether it was taken procedurally properly or not – was taken through a democratic process; and that the Contracting Governments 'ultimately control[led] the integrity of their own [C]onvention[]'.[92] This does imply that a vote would be expected and that what would be required would be at the minimum a majority vote amongst Contracting Governments – this might be difficult for Japan to achieve, given that in the two decades since 2002, the balance of numbers in the IWC has shifted in favour of the anti-whaling 'side'. However, that might not be a bad thing if it were to lead to compromise on both sides – Japan needing the support of anti-whaling Contracting Governments; and those Governments recognising and acknowledging that persuading Japan back into the IWC would be an achievement of considerable note.

New Zealand's expressed 'grave concern' over the integrity of both the IWC and all multilateral environmental agreements remains important. Reservations are controversial and complicated, but that is a descriptor of the ICRW, the IWC and whaling too. As tempting as it might be for some states to regard Iceland's re-adherence in 2002 as 'a mistake' and one not to be repeated, in the long run it would probably be the wiser course to encourage Japan to re-adhere.

Notes

1 Patricia Birnie, *International regulation of whaling: From conservation of whaling to conservation of whales and regulation of whale-watching* (Oceana 1985) 250. Norway rejoined in 1960; and the Netherlands in 1962.
2 The Seychelles, for instance, was the Contracting Government which formally proposed the formulation of the 'moratorium clause' which was adopted in 1982 – but the state is no longer a party to the ICRW. See IWC34, *Chair's Report*, 1982, 20, < https://iwc. int/private/downloads/-bZcXV5nQQeQlKLcwmXsVA/1982_Chairs_Report.pdf > last accessed 20 March 2023; or < https://archive.iwc.int/pages/search.php?search=%21col lection29604&k=# > last accessed 27 June 2023.
3 IWC59 (2007), Plenary Session Day 3, present author's personal notes.
4 IWC67, 'Chair's report of the 67th annual meeting (2022)', Annex E: 'Other Statements: Japan's statement after the vote on its proposal for the way forward of the IWC' < https://archive.iwc.int/pages/view.php?ref=7592&k=0e1e251069&search=&offset=0 &order_by=relevance&sort=DESC&archive=# > accessed 27 June 2023.
5 Ibid.
6 Ibid.
7 Of the broad pro- against anti-whaling divide.
8 The word which the ICRW uses is 'objection' – but 'reservation' is more commonly used, and would seem the better word as in general international law the word 'objection' is more commonly used to describe the response by other states to a reservation taken by a state. In this chapter, the word 'reservation' will generally be used.
9 Article 19: Formulation of Reservations.
10 Peru withdrew its objection on 22 July 1983.
11 IWC34, Verbatim Record, 1982, 79 < https://archive.iwc.int/pages/view.php?ref= 424&k= > accessed 27 June 2023.
12 IWC43, Report 1992: '25. Other business', 1993, 36, < https://archive.iwc.int/pages/ search.php?search=%21collection29604&k=# > accessed 27 June 2023.
13 Ibid.
14 Press Release, Ministry of Fisheries, Iceland, 'Government of Iceland announces withdrawal from the International Whaling Commission' (27 December 1991), < http:// www.highnorth.no/Library/Policies/National/go-of-ic.htm >; quoted in Peter Davies, 'Iceland and European Union accession – the whaling issue', (2011), *Georgetown International Law Review* 24(1), 23, 27 fn 26. [Note: The reference is to Davies – the weblink quoted in Davies appears to be broken. Another source is Keith Schneider, 'Iceland plans to withdraw from International Whaling Agreement', (28 December 1991) *New York Times*, < https://www.nytimes.com/1991/12/28/world/iceland-plans-to-withdraw-from-international-whaling-agreement.html > accessed 27 June 2023.]
15 Davies, ibid., 26–27.
16 Ibid, 27. Generally on NAMMCO see < https://nammco.no/ >.
17 Ibid.
18 Ibid.
19 IWC53, 'Chair's report of the 53rd annual meeting (2001)', (IWC 2002) 5.
20 Ibid.
21 Ibid.
22 IWC53, 'Opening statements: Iceland', IWC Archive, Cambridge, UK.

23 Ibid.
24 IWC53, 'Opening statements: Antigua and Barbuda', IWC Archive, Cambridge, UK.
25 IWC53, 'Opening statements: Solomon Islands', IWC Archive, Cambridge, UK.
26 IWC53, 'Chair's report of the 53rd annual meeting (2001)', (IWC 2002) 6.
27 Ibid., 7.
28 Ibid.
29 The Commissioner from Sweden, Bo Fernholm.
30 IWC53, 'Chair's Report', 7.
31 Ibid., 7–8.
32 Ibid., 8.
33 Sean D Murphy, 'Blocking of Iceland's effort to join whaling convention', (2002), *The American Journal of International Law* 96(3), 712, 713.
34 IWC54, 'Chair's report of the 54th annual meeting (2002)', (IWC 2003) 5–6.
35 Ibid., 6.
36 Ibid.
37 Ibid.
38 Ibid.
39 Ibid., 7.
40 Ibid.
41 Ibid.
42 For a closer consideration of the events leading up to, at, and after the 2002 Meeting, see Ed Couzens, 'Whaling and dealing: Aboriginal subsistence whaling, politics and poverty', in Yves Le Bouthillier, Miriam Alfie Cohen, Jose Juan Gonzalez Marquez, Albert Mumma and Susan Smith (eds.), *Poverty alleviation and environmental law* (The IUCN Academy of Environmental Law Series, Edward Elgar 2012), 100.
43 Ibid., 108–109; quoting from IWC54, 'Chairs' Report', 19–21.
44 IWC, 'Chair's report of the 5th special meeting', in 'Chair's report of the 55th annual meeting (2003)', (IWC, 2004) 137–148, 139. Iceland did hedge on this commitment by indicating that it would 'not apply [] in case of the so-called moratorium on whaling for commercial purposes [] not being lifted within a reasonable time after the completion of the RMS'. Ibid.
45 Ibid.
46 Ibid.
47 Ibid. The instrument stated in full: 'Notwithstanding this, the Government of Iceland will not authorise whaling for commercial purposes by Icelandic vessels before 2006 and, thereafter, will not authorise such whaling while progress is being made in negotiations within the IWC on the RMS. This does not apply, however, in case of the so-called moratorium on whaling for commercial purposes, contained in paragraph 10(e) of the Schedule not being lifted within a reasonable time after the completion of the RMS. Under no circumstances will whaling for commercial purposes be authorised without a sound scientific basis and an effective management and enforcement scheme'. Ibid.
48 Made essentially by Australia, supported by Argentina, Brazil, Mexico and New Zealand. Australia argued that a decision on competence had already been taken at both IWC53 and IWC54 – the Chair overruled this point of order.
49 By Australia and Mexico.
50 Iceland's view, supported by Antigua and Barbuda, was that the IWC does not have this competence.
51 IWC, 'Chair's Report of the 5th Special Meeting', 140–141.
52 Ibid., 141.
53 Ibid.
54 Ibid.
55 Ibid.
56 Ibid.

57 Including by Denmark, Iceland, Ireland, Norway, Sweden and the Republic of Guinea.
58 IWC, 'Chair's Report of the 5th Special Meeting', 141.
59 Ibid., 142.
60 Sweden explained its vote without explaining it – noting that 'there had been a number of procedural votes' and that 'throughout it had voted according to its legal analysis of the situation'. Sweden then made a formal declaration that Iceland's reservation 'raised serious doubts as to Iceland's commitment to the object and purpose of the [ICRW]'; and that it (Sweden) would 'consider seriously making a formal objection regarding Iceland's reservation'. Ibid.
61 Argentina, Costa Rica, Gabon, India, Kenya, Mongolia, Morocco, Oman, Panama, St Vincent and the Grenadines, Senegal, South Africa.
62 IWC, 'Chair's Report of the 5th Special Meeting', 142.
63 Ibid.
64 Ibid.
65 David Harris and Sandesh Sivakumaran, *Cases and materials on international law* (first published 1973, 9th ed., Sweet and Maxwell 2020) 670.
66 *Rawle Kennedy v Trinidad and Tobago*, Comm. 845/1999, UN Doc. A/55/40, Vol. II at 258 (HRC 1999); available at < http://www.worldcourts.com/hrc/eng/decisions/1999.11.02_Kennedy_v_Trinidad_and_Tobago.htm > accessed 20 March 2023; or OHCHR.org, 'International Covenant on Civil and Political Rights: Selected Decisions of the Human Rights Committee under the Optional Protocol', Sixty-sixth to seventy-fourth sessions, Vol 7, (July 1999–March 2002), (United Nations 2006), 5–12, < https://www.ohchr.org/sites/default/files/Documents/Publications/SDecisionsVol7en. pdf > accessed 27 June 2023.
67 See Glenn McGrory, 'Reservations by virtue? Lessons from Trinidad and Tobago's reservations to the first optional protocol', (2001), *Human Rights Quarterly* 23(3), 769–771. The reservation read: '… Trinidad and Tobago re-accedes to the Optional Protocol to the International Covenant on Civil and Political Rights with a Reservation to article 1 thereof to the effect that the Human Rights Committee shall not be competent to receive and consider communications relating to any prisoner who is under sentence of death in respect of any matter relating to his prosecution, his detention, his trial, his conviction, his sentence or the carrying out of the death sentence on him and any matter connected therewith'. *Rawle Kennedy v Trinidad and Tobago*, ibid., 4.1.
68 McGrory, ibid., 771–772.
69 Ibid., 780.
70 *Rawle Kennedy v Trinidad and Tobago* (n 66) 6.4.
71 Ibid., 6.5.
72 Ibid., 6.6.
73 It probably was the case that Trinidad and Tobago did not intend to be a Party without the reservation – see, for instance, discussion at McGrory (n 66) 815. Trinidad and Tobago subsequently withdrew altogether.
74 McGrory (n 67) 790.
75 Johanna Fournier, 'Reservations and the effective protection of human rights', (2010), *Goettingen Journal of International Law* 2(2), 437, 445–446.
76 McGrory (n 67) 811.
77 Ibid., 811.
78 Article 44(1) states that the right of a Party to 'denounce, withdraw from or suspend the operation of a treaty may be exercised only with respect to the whole treaty uncles the treaty otherwise provides or the parties otherwise agree'.
79 McGrory (n 67) 812.
80 Ibid., 825.
81 Alexander Gillespie, 'Iceland's reservation at the International Whaling Commission', (2003), EJIL 14(5), 977, 977 (abstract).

82 Anthony Aust, Modern treaty law and practice (first published 2000, 3rd ed., Cambridge University Press 2013) 142; quoted in Harris and Sivakumaran (n 64) 670.

83 It should be noted at this point that not all Contracting Governments to the ICRW agree that the ICRW is an 'environmental treaty'. That does not affect the point being made that modern treaties with environmental relevance are moving away from including provision for reservations to be lodged.

84 Gillespie (n 81) 988–989.

85 Time limits for the lodging of reservations are certainly usually respected. When, for example, in late 2014 Australia lodged a reservation to the listing of three species of thresher shark and two of hammerhead shark on Appendix II of the Convention on Migratory Species, 1979, Australia explained that it was doing so on the basis of urgency – as its national Parliament was going into a recess period, Australia felt that it did not have the time to amend its national law to remedy unintended consequences of the listing. JSCOT, 'Amendments to Appendices I and II to the Convention on the Conservation of Migratory Species' (*Australian Parliament House*, 9 November 2014) < https://www.aph.gov.au/JSCT/Report149 > accessed 20 March 2023. The point being made is merely that Australia did not believe that it had the luxury of waiting beyond the time limit prescribed by the treaty.

86 IWC, 'Chair's report of the 55th annual meeting (2003)', (IWC 2004) 6.

87 Mexico noted that it considered the procedure at the 5th Special Meeting to have been 'improper because it allowed Iceland [] to vote'.

88 IWC55, 'Chair's Report', 6.

89 IWC.int, 'Iceland and her re-adherence to the Convention after leaving in 1992' < https://iwc.int/_iceland#5 > accessed 27 June 2023.

90 Fred Attewill, 'Iceland resumes commercial whaling' (*The Guardian*, 23 October 2006) < https://www.theguardian.com/environment/2006/oct/23/whaling.conservation > accessed 27 June 2023. It should be noted, however, that while the taking of this fin whale represented a resumption of commercial whaling, between 2003 and 2007 Iceland was taking whales 'under special permit' – see Davies (n 14) 29.

91 In February 2022 Iceland did announce, through its Minister of Fisheries, Svandis Svavarsdottir, writing in Icelandic newspaper *Morgunbladid*, that commercial whaling would cease in 2024. Alex Berry, 'Iceland announces plan to end whaling in 2024' (*Deutsche Welle*, 4 February 2022) < https://www.dw.com/en/iceland-announces-plan-to-end-whaling-in-2024/a-60666862 > accessed 27 June 2023. The reason given, however, was that commercial whaling had become uneconomic. Further, on 20 June 2023, Svandís Svavarsdóttir, who is Iceland's current Minister of Food, Agriculture and Fisheries, issued a directive postponing the whaling season until 31 August 2023 – and indicated that a permanent ban is being considered; apparently after reaching the conclusion that the commercial killing of whales is not in compliance with Iceland's Animal Welfare Act. Cliff White, 'Iceland temporarily halts whale hunting, with permanent ban possible' (*SeafoodSource*, 23 June 2023) < https://www.seafoodsource.com/news/environment-sustainability/iceland-orders-temporary-halt-to-whale-hunting > accessed 27 June 2023. Whether these are indications that Iceland's commercial whaling practices will cease remains to be seen.

92 Gillespie (n 81) 998.

6 Spill-over? The Convention on International Trade in Endangered Species after Japan's Withdrawal from the Whaling Convention

Nikolas Sellheim

Introduction

In August 2019, the 18th Conference of the Parties (COP18) of the Convention on International Trade in Endangered Species of Wild Fauna and Flora (CITES) took place in the Geneva after having been rescheduled and moved from Sri Lanka to Switzerland due to recent terror attacks at that time. Similar to the IWC, the CITES has been struggling with different 'camps', i.e. with a rift that does, on many occasions, not allow for consensus-based decision-making, for a long time. Especially the cases of whales and elephants that are placed on Appendix I and Appendix II of the Convention has caused problems of agreement amongst the parties.[1] However, a modus operandi has been found at COP10 in 1997 in so far that stable elephant populations were placed on Appendix II—meaning, regulated international trade in them is possible under international oversight—while those populations that were, and are, still vulnerable or endangered have remained on Appendix I. This listing principally prohibits international trade in elephant specimens unless for very specific purposes. While this 'split-listing' is possible, it should, according to CITES Resolution Conf. 9.24 (Rev. CoP17), be avoided.[2]

After Japan had withdrawn from the ICRW in July 2019, many governments, e.g. Australia, have condemned Japan for this move.[3] While that may be so, it remains rather unclear why governments have criticised Japan since it has ever since its withdrawal given up on the scientific 'special permit' (scientific) hunts permitted under Article VIII of the International Convention for the Regulation of Whaling (ICRW) in the Antarctic and has therefore shifted its whaling operations into its own territorial waters and its exclusive economic zone (EEZ). The international effects of Japan's whaling have consequently decreased significantly, even strengthening whale conservation efforts in the Southern Ocean Whale Sanctuary (SOWS) and making an Australian whale sanctuary possible in the near future.[4]

Other states, however, responded differently to Japan's move—either purposefully or subconsciously. The Namibian Minister of Environment and Tourism, Pohamba Shifeta, for example, announced at CITES COP18 that it, along with other countries of the South African Development Coalition (SADC), would consider withdrawing from CITES. Background for this announcement was that the COP

DOI: 10.4324/9781003250814-7

voted against the downlisting of white rhinos from Appendix I to Appendix II. Since Namibia's white rhino population is one of the largest in the world, the country wishes to open up limited international trade in white rhino specimens. However, since the COP voted against this proposal, Shifeta and his colleague from Botswana, Kitso Mokaila, shared their resentment with the international press, also by expressing their country's potential withdrawal from CITES if the trade restrictions for their respective white rhino populations are not lifted.[5]

The impression one might easily get is that when a treaty does not suit a state's interests anymore, the state will leave. Indeed, CITES Article XXIV also allows for a Party's withdrawal "by written notification to the Depositary Government at any time. The denunciation shall take effect twelve months after the Depositary Government has received the notification." Most treaties have some exiting clause, enabling parties to exercise their sovereign right to be (or not to be) party to a treaty.[6] At the same time, however, once ratified, a treaty's provisions are embedded in the domestic legal system, which, in turn, would have to be changed after withdrawal, bearing its own difficulties.[7]

In this chapter, I will explore the likelihood of other conventions potentially going through a similar fate as the IWC with parties leaving if they are no longer content with the majority opinion. I look primarily at CITES, but will also take into account recent developments under the Convention on Conservation of Migratory Species of Wild Animals (CMS or Bonn Convention). In order to avoid any potential spill-over effect onto other conventions, I propose a different mode of conservation governance, based on Bonebrake et al. (2019),[8] which will be outlined in more detail below.

Controversy within CITES

CITES was adopted in 1973 and has a current membership of 184. That means that the Convention has found almost global recognition. Currently, more than 38,700 species are listed on the Appendices. Apart from the two Appendices referred to above, a third Appendix exists that lists species that are protected nationally but for which a Party requests support from others. Of these species, the vast majority are flora species listed on Appendix II (32,364 species), while 5,056 species plus 15 subspecies are fauna species. 687 animal species, plus 32 subpopulations are included in Appendix I, compared to 395 plant species plus 4 subpopulations. It is worth noting that of the 687 animal species included in Appendix I, almost half are mammals.[9]

The *raison d'être* of the Convention is, as the name implies, the protection of wildlife through the regulation of international trade. That allows to infer that in some cases, it is international trade that significantly contributes to the decline in biodiversity. This would imply that scientific data informs the decision-making process in so far as it is determined in how far international trade impacts the conservation status of a species. Consequently, Resolution Conf. 9.24 (Rev. CoP17) on amendments to the Appendices refers to biological *and* trade criteria which allow for amendments to the Appendices (which are to be decided by the COP with a 2/3 majority).

As Challender and MacMillan (2019) demonstrate, however, 'listing decisions are frequently characterised by heated and polemic debate [...] and in reality are made for political, economic, philosophical, and even emotional reasons, as well as scientific reasons.'[10] This means that despite the existence of certain criteria, CITES cannot be considered primarily a science-based organisation, especially since it still remains a matter of debate whether international trade has ever been the sole, or at least primary, reason for declines in wildlife populations. In fact, disagreement exists over the question whether the inclusion of species in the CITES Appendices has actually been beneficial for the species.[11] Either way, while the inclusion of a species in Appendix II (and possibly even Appendix I) is made rather easy because of the application of a stringent form of the precautionary principle,[12] a downlisting is significantly more difficult, as the example of African elephants demonstrates. Especially when looking at so-called 'charismatic megafauna', a downlisting appears ever more difficult. One of the reasons might also be the presence of very large numbers of non-governmental organisations (NGOs) that appear to present the will of society at the meetings. As Figure 6.1 shows, for around half of the COPs the number of NGOs exceeded the number of parties that were present. Only since COP 12 in Chile in 2002 this trend has been reversed, even though the number of NGOs has been increasing again since then.

Given that NGOs exert pressure on national governments, the government of St. Vincent and the Grenadines proposed the adoption of a Code of Responsibility at the 69th meeting of the CITES Standing Committee in 2017. Since CITES has its own Rules of Procedure based on which NGOs can participate and intervene at CITES meetings, the proposal was rejected and deemed unnecessary.[13] This demonstrates how differently Parties to the Convention perceive the existing system, particularly in relation to NGOs.

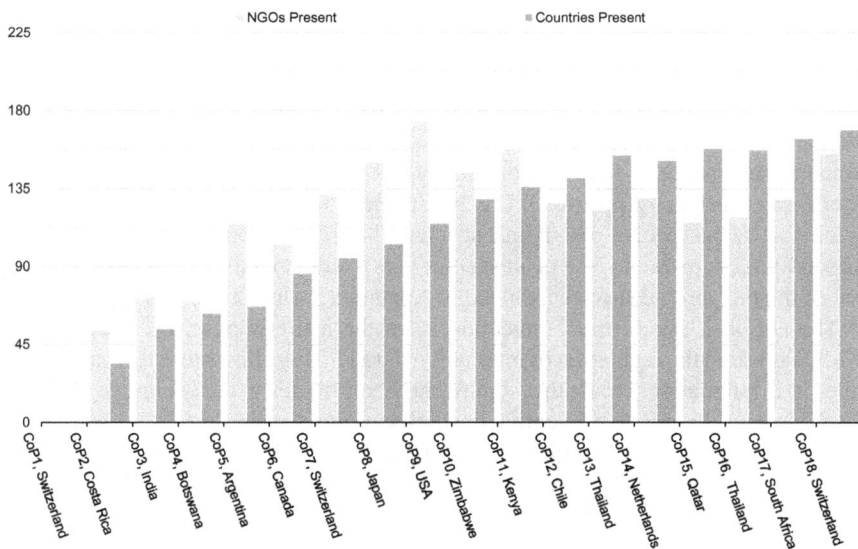

Figure 6.1 NGOs present at CITES COPs.

Another matter of potential controversy is the relationship to the IWC. From the very early days on, those whales that have been under the ambit of the IWC were automatically listed under Appendix II of the Convention. In the late 1970s, when the IWC placed more and more whale species under full protection from commercial whaling, also CITES responded by placing those species onto Appendix I. When the full moratorium on commercial whaling was adopted by the IWC, it was these species for which no export or import permits were issued anymore, essentially placing them on Appendix I as well.

Over time, the 'normal' mode of whales not being able to be traded in anymore has taken hold of the Convention and its *modus operandi*. However, the question has been asked whether or not there is a primacy of the ICRW over CITES, identifying it as *lex superior* under international law. One of the arguments that can be brought forth in this context is the mere fact that the ICRW was adopted already in 1946, while CITES was adopted only in 1973. If the two regimes had the same scope, e.g. if *both* were to deal with whaling, then the latter regime, *lex posterior*, would take hold. However, since the mandate of CITES is significantly different to that of the ICRW, albeit them being closely interlinked, CITES becomes a regime that follows the lead of the ICRW and with that of the IWC.[14]

However, CITES Parties have repeatedly criticised the fact that the moratorium has led to a quasi-automatic inclusion of many whale species in Appendix I. Both Japan and Canada (Canada left the IWC shortly after the moratorium was put in place) noted at COP9 in 1994, for example, that the processes and criteria were fundamentally different between the IWC and CITES, and that a lifting of the moratorium would still prevent CITES Parties from engaging in trade in whale products due to the species' listing in Appendix I.[15] Still, the majority of states recognise the primacy of the IWC, especially since Article XV of the Convention in the context of marine species requires the Secretariat to consult with 'intergovernmental bodies having a function in relation to those species', while Article XIV relieves CITES Parties from the provisions of the Convention in case they are party to an agreement that deals with the conservation of marine species. While the IWC is not explicitly mentioned in the text of the Convention, at the time of its conclusion it was the only body for the protection of whales, which is an indication of its dominant function in the early 1970s.[16]

Against this backdrop, let us now turn back to the question this chapter seeks to address: Is it likely that Japan's withdrawal from the ICRW has implications for other multilateral environmental agreements (MEAs) as well?

The Root of the Problem

In my view, the root of the problem lies in the fragmentation of international law. In the absence of a global, general law-making body, different spheres of international law, such as human rights law, environmental law or law of the sea, have developed over time, each with its own processes, mechanisms and institutions. Even though that is so, this does not mean that each of the different branches exists in isolation from the other. Much to the contrary: they frequently interact, overlap, interplay and refer to one another.[17]

That also means, however, that each of the regimes in question has its own Rules of Procedure, which enables the parties to engage in their own decision-making without being dictated what to do by other regimes. The case of the IWC-CITES interplay has shown though that state parties can voluntarily adhere to the decisions of another regime or simply follow its lead. CITES has made rather unmistakably clear that it considers itself as a prolongation of the IWC with regard to international trade in whale products.

With Japan having left the IWC, both this interplay and the effects of such withdrawal are rising to the surface again as it puts the strong link into question. Japan, as the spearhead of whaling nations, can consequently ask why it should be bound to a body's decision not to engage in commercial whaling even though it has 1. consistently worked against that decision; 2. is not even part of the organisation anymore; 3. is interested in the international trade in whale products, and 4. has argued—and arguably demonstrated—that the moratorium is inherently unscientific. And apart from these points, Japan has furthermore engaged in scientific whaling in the Antarctic as part of its different whaling programmes—a fact that many have coined a 'loophole' in the moratorium[18]—and has introduced whale products from the sea, which falls under the aegis of CITES and the requirements for Appendix I and II listings.

By having left the IWC, Japan is, in the end, better off than before. On the one hand, it is now able to engage in whaling to a degree that it deems necessary and sustainable, based on its own scientific research. Since it is still an observer to the IWC and still collaborates closely with the IWC's Scientific Committee—e.g. by providing and obtaining data on conservation statuses—while also being an observer to the North Atlantic Marine Mammal Commission, it does not act in isolation. And this, in turn, is a requirement for the conservation of marine mammals, and especially cetaceans, under the UN Convention on the Law of the Sea, Article 65. This article sets out that for the conservation of cetaceans states should cooperate through 'appropriate international organisations'. Since the article does not contain any more specifics, Japan can be said to fulfil this requirement.

On the other hand, Japan is much better off after having left the IWC since it is no longer the public scapegoat for whaling. Even though Japanese whaling is still very much opposed by states and organisations worldwide, Japan does not conduct whaling anymore in the Southern Ocean Whale Sanctuary, thereby not potentially infringing upon the rules and regulations set by the IWC; and second, since it is no longer a member of the IWC, the spotlight of international media attention is in all likelihood not focused on Japanese whaling, at least to a lesser degree than before. The lack of media attention regarding the last meeting of the IWC in 2022 underlines this fact.

This allows for the conclusion that Japan's withdrawal will probably not result in a re-adherence to the convention in the near future. The conditions for Japan are consequently significantly different to those of Iceland when it re-adhered to the ICRW in 2002.[19] The benefits to leave the IWC therefore outweighed the benefits of staying in the Commission. Within CITES, it is now potentially able to steer a different course than before.

Another issue that affects the satisfaction of a state within CITES is the fact that all Parties have the same voting powers, irrespective of the species or population. Meaning, if an endangered species occurs in, say, two African states and these two states actively trade in it, a European state can table a proposal to include this species in one of the Appendices. Ultimately, this proposal would be put to a vote by the COP. If successful, states that do not constitute a habitat for the species make an outside decision that would significantly affect the two African states in question, also neglecting any conservation initiatives they would have in place. After all, there are no hierarchical rules for the amendments of the Appendices in the sense as those states that constitute a species' habitat would have more decision-making powers than those that do not. The premise of 'one state, one vote' is a fundamental element in international law. Any state can therefore table a proposal for amendments of the Appendices, based on biological and trade criteria stipulated in Resolution Conf. 9.24 (Rev. CoP17). The stringent nature of the precautionary principle simplifies this process. The proposal is then voted upon and decided by a simple two thirds majority of all Parties, as stipulated in Article XV.1 (b).

In other bodies, this majority vote is circumvented and/or based on specific criteria. Within the Council of the European Union, for example, a so-called qualified majority voting procedure has been established when votes are taken. This qualified majority is reached when, first, 55% of the EU Member States vote in favour of a proposal and, as a second condition, representing at least 65% of the total EU population.[20] In the interest of fairness within CITES, a similar process might arguably be beneficial. This would require an amendment to the Convention, however, which, again, two thirds of the Parties would have to agree to as per Article XVII—a scenario that is highly unlikely as it would fundamentally alter its *modus operandi*.

The Choice for or against a Regime

For the purposes of this chapter, I consider two elements that are essential for a state to join a certain regime or to position itself against it, either by not joining at all or by leaving it. On the one hand, a rational choice is being made by the state. The equation is quite simply put forward by Guzman (2008):

> [S]tates will only enter into agreements when doing so makes them (or, at least, their policy-makers) better off. [...] Treaties allow them to resolve problems of cooperation, to commit to a particular course of conduct, and to gain assurances regarding what other states will do in the future.[21]

The flip-side here is, of course, that when a state, or its policy-makers, consider they are worse off, and that cooperation is no longer guaranteed and does not resolve state-specific problems, it no longer can be justified for the state to remain in the regime. The other issue relates to the legitimacy of a regime. Generally speaking, a regime, and especially an environmental regime, was put in place in order to correspond to one or more specific environmental problems that require social change. This means, in principle multilateral environmental agreements (MEAs) are essentially problem-driven and should contribute to problem-solving to achieve a certain goal. While there are numerous definitions of 'legitimacy' in international law and politics,[22] for the purposes of this study I refer to Breitmeiter (2008),[23] who notes that regimes deserve following if:

1   they contribute to a reduction of uncertainties in the knowledge-base
2   they achieve compliance with regime norms
3   they have a causal impact on goal-attainment and environmental problem-solving
4   their policies reflect normative claims of equity and sustainability and
5   procedures exist that ensure the participation of global civil society.

A crucial issue in this context is the question of the degree of 'output legitimacy' and 'input legitimacy'. Output legitimacy refers to the performance and effectiveness of a regime relating to the outcomes of cooperation. Input legitimacy refers to the inner ways a regime functions, for example issues related to equity and the will of the people or member states.[24]

For Japan leaving the IWC, it was a mix of both that has prompted the decision: on the one hand, the goal of the whaling convention is considered differently by the IWC member states, which inevitably leads to uncertainties of what the regime is actually to achieve—hence, the degree of output legitimacy depends on the way the ICRW is interpreted.[25] On the other hand, Japan has long criticised the way the IWC has been dealing with the moratorium on commercial whaling, which it considers inherently unscientific and hence without any justification.[26] Any Japanese attempt to change the status quo has thus far failed. Consequently, the degree of input legitimacy—from a Japanese perspective—has been low since it considers the ICRW a convention for the management of whales and whaling and not a convention for the preservation of whales. Since no change in the status quo of the IWC can be expected in the near future, from a Japanese perspective leaving the IWC appeared very reasonable in light of the criteria for legitimacy outlined above.

CITES, however, has been hailed as one of the most successful conservation regimes in the world. With a membership of 184, it is indeed one of the most ratified environmental conventions. Contrary to the IWC, CITES Parties do not disagree over the object and purpose of the Convention, namely the conservation of biodiversity through the regulation of international trade. In that sense, output and input legitimacy appear to be high for this convention. The question is why CITES has experienced such a high degree of approval from the vast majority of UN Member States since its inception?

As Barrett (2003) shows, there are many different reasons for states to join a treaty. Generally, when states join a treaty, the benefits of membership outweigh the costs of not joining. When a treaty or agreement, in addition, contributes to rectifying an environmental problem, it can, without a doubt, be considered successful.[27] The Montreal Protocol having contributed to the drastic slow-down of ozone depletion is a case in point in this regard. The Kyoto Protocol, however, is a regime that has not experienced this high degree of approval as the Montreal Protocol. Especially the non-adherence of some of the main polluters, such as the United States, weakens the efficacy of this regime. Its successor, the Paris Agreement, whilst having a membership of 193 states—thus indicating a unified, global wish to tackle the problem of global warming—is likely failing to reach its ultimate goal: to limit global temperature rise to 2°C, but preferably to 1.5°C compared to pre-industrial levels. As the US's National Oceanic and Atmospheric Administration (NOAA) shows, in 2021 the average temperature on the Northern Hemisphere was 1.1°C higher than in pre-industrial levels, with no trend change in sight.[28]

A similar argument can be made about CITES: despite its quasi-global membership, it has obviously not contributed to a slow-down in biodiversity loss, as the latest report of the Intergovernmental Platform on Biodiversity and Ecosystem Services (IPBES) has so dramatically shown.[29] Instead, biodiversity loss continues unabated. It is therefore rather provocative, but somewhat justified to say, that CITES, along with all other biodiversity-related conventions, has failed. Still, the vast majority of countries still eagerly participates in the treaties, since, even though the biodiversity crisis is reaching dangerous levels, international cooperation is crucial for current and future strategies to combat biodiversity loss. One state cannot do it alone. This consequently justifies the choice to join and adhere to biodiversity-related conventions.

At the time of CITES' inception, the dramatic loss in all levels of biodiversity was not yet clear and could not have been predicted. In the case of whales and polar bears, for example, and the approximated 1,000 species that were listed on the Appendices in the early 1970s, international trade, and not the multifaceted threats of today, posed an important threat to the survival of a species. The international trade-environment in biodiversity, however, has changed over the years and decades. At this point in time, especially against the backdrop of the discussions on international wildlife trade and the outbreak of zoonotic diseases, international trade in biodiversity from the wild does not play such a prominent role anymore as it did in the 1960s and 1970s. Especially in the case of whales, hardly any international trade has been conducted over the last few years—the exception being a very limited trade in fin and minke whale meat between Iceland, Norway and Japan.[30] Not surprisingly, previously strongly overhunted species, e.g. the humpback whale, have reached sustainable and growing levels.

One might argue that there are essentially three reasons for rebounding whale populations: first, the moratorium on commercial whaling is still in place and has therefore contributed to lesser animals being whaled; second, despite the existence of some commercial (and scientific) whaling, the scale of these activities has never reached industrial levels and the numbers taken were (and are) much lower than in the mid-20th century; third, the international *zeitgeist* on whaling has changed dramatically with only very few countries still actually engaging in commercial whaling. A fourth reason could be mentioned in this context as well: the commercial whale hunt currently undertaken does hardly have any international dimensions. Norway and Japan primarily hunt whales for their domestic markets, while merely

Iceland's fin whale meat export to Japan was international. In that sense, CITES has no longer any justification to keep all whales protected by the IWC moratorium in Appendix I because neither the biological nor trade criteria justify this. CITES is therefore not elementary for protecting whale species. This, of course, is subject to debate.

Why do countries remain in CITES? As mentioned above, the benefits obviously outweigh the costs. How seriously can it then be taken when Namibia's Environment Minister threatens to leave the Convention? After all, Japan has shown that leaving a controversial treaty may be more beneficial than staying in it.

Namibia and CITES

Namibia joined CITES in December 1990, with the Convention entering into force for the country in March 1991.[31] The decision to join the Convention was taken in the same year as Namibia gained independence from South Africa. It cannot be ascertained why Namibia joined CITES in the first place, but it may be possible that it was particularly the international trade in ivory which prompted newly independent Namibia to join, simply because it was anticipated that CITES membership might ease access to international ivory markets. This, however, remains in the realm of speculation. Another reason may be that Namibia has from the very beginnings of its CITES membership pushed for less stringent prohibitions. In other words, it may have joined CITES to make international trade in some endangered species possible again.

The reason for this speculation is that at its first COP in 1992 (COP8), Namibia has tabled or co-sponsored eight proposals that aimed at downlisting certain animal species from Appendix I to Appendix II or completely deleting them from the Appendices. Concerning the former, it immediately aimed at one of the most iconic species: the African elephant (*Loxodonta africana*) to be moved from Appendix I to Appendix II. Other charismatic species, such as the Aardvark (*Orycteropus afer*) or the brown hyaena (*Hyaena brunnea*), were to be deleted from the Appendices completely.[32] In the case of the Aardvark, Namibia succeeded.

It almost appears as if Namibia's first attempts served as a trial-run for its apparently most important goal: the downlisting of its elephant populations. While at COP9 no Namibian proposals were tabled, at COP10 it fully focused on the downlisting of elephants. After heated debates, the elephant populations of Namibia, Botswana and Zimbabwe were transferred from Appendix I to Appendix II.[33] In the COPs to come, Namibia primarily focused on amending the annotation that is part of the Appendix II-listing of its elephant population. At no other COP did the country table or co-sponsor proposals for as many species as at COP8. Instead, virtually at every COP, its elephant population played a prominent role. Merely on two occasions did it support the transfer or deletion of species: at COP11, the deletion of the brown hyaena from Appendix II, co-sponsored by Switzerland, and at COP13 the deletion of the rosy-faced lovebird (*Agapornis roseicollis*) from Appendix II, co-sponsored by the United States. But almost at every single COP, elephant issues were raised.

This changed at COP18 in 2019—the meeting where the Minister announced a possible withdrawal. While again the annotation to its elephant population was raised as an agenda item, now the Southern white rhinoceros (*Ceratotherium simum simum*) had shifted into the focus of the country. As was the case with the elephants, Namibia proposed to downlist its white rhino populations from Appendix I to Appendix II and thus to achieve a split-listing. Not surprisingly, this attempt failed.

The successful downlisting of its elephant population and the attempted downlisting of its white rhino population appear to be part of a larger narrative that Namibia pursues. Even though the Appendices play a major role in the Convention, the discussions on the floor deal with important issues relevant for wildlife conservation. In this regard, Namibia has since COP9 become the spearhead for a movement that links trophy hunting with the benefits for local communities and thus for the benefits for biodiversity. While this issue has not been put forward by Namibia at every COP as an agenda item, the country nevertheless establishes and defends this link in the discussions on different matters. The underlying, prevailing narrative is that 'hunting is good for conservation'—a narrative that many so-called sustainable use countries make use of.

At COP17 in 2016, it became once more obvious that Namibia pursues an agenda that aims at a reopening of the ivory trade. Along with Zimbabwe and South Africa, Namibia handed in a proposal that dealt with the decision-making process related to trade in ivory.[34] And at COP18, Namibia, along with Botswana, Democratic Republic of Congo, South Africa and Zimbabwe handed in a proposal that aimed to establish a Rural Communities Committee in order to enable rural communities to engage better in the decision-making processes related to CITES that affect them.[35] The need for such a committee was reaffirmed by Zimbabwe's Minister for Environment, Climate, Tourism and Hospitality Industry in June 2022, even though he notes that at this time, leaving CITES is not yet an option.[36]

In addition to the above, at COP19 Namibia co-sponsored a proposal to amend the criteria by which the Appendices are amended. Along with Botswana, Cambodia, Eswatini and Zimbabwe, the country aimed to include livelihoods criteria into Resolution Conf. 9.24 (Rev. CoP17) before a species is proposed to be included in the Appendices.[37] Since this would mean a significant change to CITES, the proposal failed.

These documents taken together, paired with Namibia's past and consistent focus on elephants and, since COP19, white rhinos make it appear unlikely that it would actually withdraw from the Convention since Namibia works hard on the way CITES should function (in the future). While it appears reasonable to assume that in light of the rather steadfast positions of CITES Parties on these matters, there is no benefit concerning these species to remain in the Convention, the issue of white rhinos is new and has not been part of Namibia's ambitions before. Moreover, related to elephants and trade in ivory, the country has experienced setbacks before—whilst having been successful in downlisting the elephant populations—and to use the new topic of white rhinos to threaten with a withdrawal from the Convention appears to be motivated by domestic politics. After all, while COP19 took place in August 2019, the Namibian general elections were held in November of the same year.[38] Namibia's Environment Minister was consequently in the midst of an election campaign, demonstrating to the Namibian people that he appears to stand up for the interests of rural communities by enabling them to benefit from rhino trophy hunting and the export of live animals.

What the Minister did not mention is that Namibia benefits greatly from CITES. Especially in regard to the exchange of information and the close cooperation with other Parties, poaching and the illegal trade in wildlife can be tackled more effectively. Moreover, Namibia hosts a large number of species that are listed on the Appendices: at present, based on the Species+-list produced by United Nations Environmental Programme—World Conservation Monitoring Centre (UNEP-WCMC) and the CITES Secretariat, currently 301 species on the Appendices are listed as being located in Namibia.[39] Apart from white rhino and African elephant, a vast number of endangered plant species benefit from international cooperation between Namibia and other CITES Parties. What he also did not mention is the fact that the equation of trophy hunting immediately benefiting rural communities is simplified at best. Stasja Koot, for example, shows that while economics and science may support trophy hunting, the social realities of trophy hunting in Namibia go beyond mere 'benefit'-considerations but must be considered in larger frameworks of social interrelations within a certain region. He argues that while communities may benefit from trophy hunting economically, thus making it a 'success' and an easy election-narrative, it also contributes to the erosion of traditional social interrelations and to the dissolution of traditional human-wildlife interactions. Through trophy hunting, he argues, wildlife is being commodified and blent into a neoliberal perception of the environment,[40] inevitably making it yet another tool within wider discourses of colonisation.[41] Moreover, while communities may economically benefit from trophy hunting, it also makes them dependent, first, on the operators of hunting salaries and, second, on this stream of income. Tour operators are therefore able to pay rather small salaries and exploit the lack of other employment as a justification for this.[42] Additionally, if the economic benefits from trophy hunting decrease, the entire community and resource management systems might be affected.[43]

The case of Namibia within CITES is therefore significantly different to that of Japan within the IWC. In Japan's case, the country has for approximately 40 years struggled to have the moratorium lifted with continuous setbacks despite providing new proposals. In other words, Japan has not only tried for a long time to cooperate with others concerning whaling but also to have its own whaling interests recognised. This has led to an impasse in the IWC where no solution could be found—an issue that was officially stated in 2006 by the Revised Management Scheme (RMS) Working Group.[44] Within Namibia and CITES, no such impasse can be noted. Of course, Namibia's proposal to downlist its white rhino population was rejected, but this cannot be compared to the major differences that have existed between Japan and the majority of IWC members over fundamental issues of whether or not whaling can be permitted. A logical result, after decades of negotiations, was for Japan to withdraw from the Whaling Convention.

Another point that speaks against a serious threat of Namibia leaving CITES is the fact that it would then be a non-Party. While Article X of the Convention specifies that trade with non-Parties is also possible, CITES Parties should still require the same documentation as from Parties from them. Essentially, therefore, nothing would change for Namibia and its trade ambitions. This fact is further stipulated in Resolution Conf. 9.5 (Rev. CoP16) on Trade with States not party to the Convention.[45] Apart from not being bound to the provisions of the Convention itself, Namibia would still have to deliver all relevant documentation. After all, its primary trading partners would still be Parties to the Convention.

Can Other Treaties Be Affected? A Glimpse at the Convention on Migratory Species

Japan having left the IWC is a rather radical step. After all, the IWC is a prominent body discursively and surfaces in media and other reporting frequently as *the* international body for whale management and conservation. Does Japan's behaviour therefore have impacts on other regimes? The most obvious one would have been CITES, particularly in light of the statement of Namibia's Environment Minister. However, as we have seen above, this does not appear to be more than rhetoric. How about the Convention on Migratory Species (CMS or Bonn Convention)? What developments can be observed that may be related to Japan's withdrawal?

At present, the CMS has a membership of 133 states, particularly from South America, Europe, Africa, South-Central Asia and Australia and Polynesia. The United States, Canada, Russia, Japan and China, i.e. some of the most important countries with great landmasses, are not party to the convention.[46] Similar to CITES, the Bonn Convention also works through two Appendices and, again similar to CITES, is species-based. That means that species listed on Appendix I are to be fully protected unless for very special purposes (e.g. the traditional utilisation of indigenous peoples), while for species listed on Appendix II special, stand-alone Agreements are to be concluded with their own membership. Currently, 657 species are listed on both Appendices of which around 200 are listed on Appendix II thus theoretically requiring the existence of around 200 separate Agreements (or Memoranda of Understandings, MoUs). Currently there are seven legally binding Agreements and 19 MoUs.[47]

Contrary to the Whaling Convention and CITES, the CMS has never had as much divergences. Since its inception in 1979, decisions have been taken by consensus and while there were naturally disagreements over the inclusion of certain species in the Appendices—especially Appendix I—these could be rectified to such a degree that consensus-decisions could be taken. As the records of the convention show, over the years the tone has sharpened to some degree. At COP12 in 2018, the first indication of a changing negotiation environment reared its head over disagreements on the inclusion of chimpanzees (*Pan troglodytes*) in Appendix I and II. Uganda tried to exclude its own population from a listing since it did not consider it 'migratory' in the spirit of Article I.1.a. Also, the inclusion of the African lion (*Panthera leo*) in Appendix II was opposed by Uganda, Tanzania and South Africa for the same reason. And a third disagreement occurred over the inclusion of the cheetah (*Panthera pardus*) in Appendix II, which South Africa, Zimbabwe and Tanzania opposed again for the same reason. Despite the convention's history of consensus, COP12 saw the first majority votes in the history of the Bonn Convention—which were cast against the opposing states.

This trend continued at COP13 in 2020 where two votes were held over the inclusion of the smooth hammerhead (*Sphyrna zygaena*) and the school shark (*Galeorhinus galeus*) in Appendix II. This time, it was Australia and New Zealand that were unhappy over the inclusion as they did not consider the sharks 'migratory'. Again, the opponents were outvoted and the species were included in Appendix II. What is noteworthy in the context of these votes is, first, that votes had to take place in the first place and that consensus-building was not possible. Second, disagreements did not arise over the conservation status of the species in question, i.e. the science was not put to the test, but rather about the interpretation of a key term of the convention: migratory. This is reminiscent of the disputes in the IWC over the object and purpose of the Whaling Convention. Even though in the CMS the object and purpose is rather clear ('the importance of migratory species being conserved [...]' and 'the need to take action to avoid any migratory species becoming endangered; ' CMS, Article II.1 and II.2), it is nevertheless the application of key terminology which leads to dispute that break with the tradition of finding consensus.

Since the above trend is rather new, it is too early to predict which direction the COPs of the Convention will go. But in light of the above, Japan's withdrawal from the ICRW and Namibia's and Botswana's announcement to withdraw from CITES, it has become clear that the tone in international conservation law has sharpened. As a consequence, one might be easily inclined to think that countries that are unhappy with the developments in a convention might leave it if difficulties persist. This might especially be the case of a convention, like the CMS, which has not found global recognition and which is being accused of not having sufficient 'teeth' to make it effective.[48] But here the question is what to lose and what to gain from staying in a convention. At present, the chronically underfunded CMS still appears to be beneficial for its Parties even though there are disagreements and votes need to take place. If, however, a country faces these votes time and again, and it is time and again outvoted, it might potentially become increasingly unhappy with the path the convention is treading. And then a process might start to fundamentally reconsider the status of the convention for the Party. Japan's constant struggles in the IWC which lasted for almost 40 years and its ultimate withdrawal from the ICRW may indeed serve as a reference point—if not legally, but at least politically.

Why exactly it is only in recent years that votes have taken place cannot be ascertained at this point. It seems though that the opposing states did not want to subject 'their' populations to the regulatory mechanisms of the convention. That is, they played the sovereignty-card and therefore consider themselves as the best protectors of the respective species. Otherwise they might not have opposed the inclusion into Appendix II.

Transformative Regime Change As a Means to Prevent Spill-over?

Even though the likelihood of Japan's withdrawal producing a spill-over effect for CITES or the Bonn Convention remains low, the case of the SADC countries has shown that the discourse on the benefit or disadvantages of remaining within these is well underway. Paired with the constant decline in biodiversity, this begs the question whether the current state of affairs and the functioning of MEAs is proficient enough to solve the apparent environmental problems and to satisfy all Parties as equal entities in international law.

Enters Bonebrake et al.'s paper *Integrating Proximal and Horizon Threats to Biodiversity for Conservation.*[49] In this paper, the authors assert that there is no 'major' threat to biodiversity, but that there is a multitude of threats operating at different spatial and temporal scales. Especially so-called horizon threats, the most obvious of which is anthropogenic climate change, can hardly be tackled by institutions that primarily deal with immediate, short-term threats such as overexploitation. While these threats are undoubtedly interlinked, they can, and should, be tackled differently and by different institutions, now commonly referred to as transformative regime change (TRC) (see Figure 6.2.).

That said, species are subject to different threats in different locales and no threat affects species equally across their range. This means that species populations show differences in their resilience towards specific threats while they are exposed to them on different levels.[50]

The IWC and CITES as well as the Bonn Convention have all integrated the multitude of threats into their working procedures even though they might not be fit to tackle these threats. After all, the IWC was founded to regulate whaling, CITES was put in place to slow down species decline through the regulation of international trade and the Bonn Convention was adopted to protect migratory species from overexploitation. None of these regimes intended to address large-scale threats such as pollution or horizon threats such as climate change.[51]

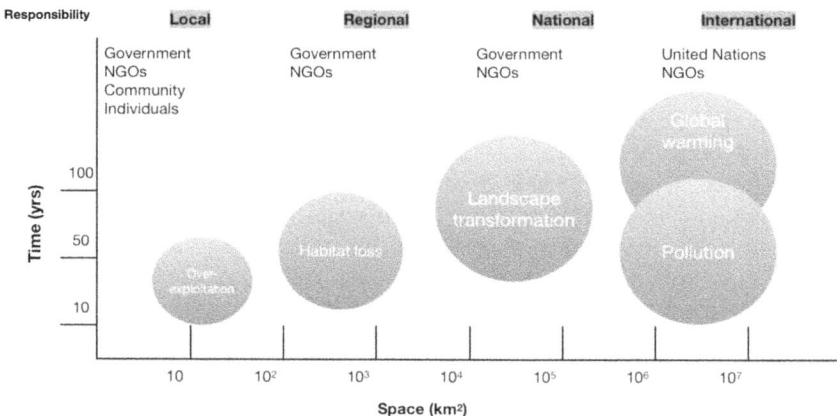

Figure 6.2 Space–time template to identify proximal and horizon threats to biodiversity.

Japan's withdrawal from the IWC was rooted in the conviction that limited and strictly controlled whaling was possible and would not affect the positive conservation status of some species. This means that Japan has consistently argued that it is the conservation of whales for purposes of maintaining the whaling industry that stands at the core of the Commission's work. Others, however, object to this view since they consider the conservation of whales as the primary objective of the Commission.[52] Japan's withdrawal therefore aims to point to the fact that the original purpose of the ICRW has been exceeded by the contemporary operations of the Commission since it now addresses and tackles issues that were not the original source for its establishment.

A similar argument could be made concerning CITES: if international trade is not a major factor for species decline, why place the species on the Appendices? This question could also be turned into a recommendatory statement: 'If it can be shown, projected or inferred that trade is not a major driver for species decline, states should consider removing the species from the Appendices'. For this to happen, the criteria for the amendments to the Appendices should be changed and be approved by the COP. Botswana, Cambodia, Eswatini, Namibia and Zimbabwe have tabled a proposal to change these criteria for COP19, as mentioned above.[53] The proposal aims to make it compulsory to table a proposal for a species' inclusion in Appendix I *only* when international trade has been identified as the major threat. Moreover, the proposal aims to make the local population part and parcel of the decision-making process for amending the Appendices.

Whether a proposal such as this is likely to ever be accepted cannot be ascertained. Yet, *if* such a change were to occur, CITES would certainly stick to its own mandate and focus solely on the threat of trade to endangered species. Such clarity in focus would consequently leave other threats such as pollution, habitat or climate change to other institutions and enable CITES to concentrate its efforts on the link between endangered species and international trade. This would reduce costs of enforcement and possibly prevent states that have a successful conservation outcome to become increasingly unhappy with CITES listing-decision.

Contrary to the IWC, CITES still has the opportunity to avert a development that is marked by fundamental disagreements, potentially even leading to a deadlock. The interpretation of the treaty itself leaves rather little leeway and it is clear that the link between endangered species and international trade constitutes the heart of the Convention. Therefore, addressing the immediate threat that is caused by international trade without including other threats enables a channelling of resources, making the treaty potentially more effective and preventing other threats to biodiversity. Regarding the latter, imagine a local community dependent on the trade in a local Appendix I-listed plant species. If this trade is now suspended, the community might be forced to relocate or to dissolve, thereby contributing to a long-term change in the local social structure and potentially increasing pollution and overexploitation of other species in the long run. Or to turn this around: suspending the legal international trade in endangered species does little to address issues other than biodiversity loss caused by this trade since sustainable livelihoods dependent on this trade are in fact considered crucial for the protection of biodiversity.[54]

But even if CITES or the IWC would solely focus on their respective objectives, they would not act in isolation. After all, CITES' Strategic Vision 2021–2030[55] places great emphasis on collaboration with other international environmental agreements, NGOs, international and regional partners. If focusing solely on trade, it would, however, leave issues that go beyond international trade to exactly these partners. Also the IWC and CITES collaborate closely through various resolutions and decisions on both sides. Yet, by acknowledging long-term or even horizon threats, but still focusing on their own original mandate, no dissolution of foci would occur and each institution would benefit from the findings of the other without having to potentially replicate the work that has been carried out by others.

To prevent any potential for spill-over of Japan's move to leave the IWC, it is therefore imperative that the focus rests on the specific threat a regime aims to address. In the case of CITES, unsustainable trade in endangered species and in the case of the Bonn Convention, unsustainable exploitation of migratory species was the key element for their coming into being. Consistent enforcement of the provisions of the respective conventions with a clear focus on what they aim to achieve can consequently be considered a key element for making them work, provided, of course, that compliance occurs throughout—an issue which is also of dire concern for CITES.[56] This means that the responsibility to address and tackle certain threats is being relegated to different fora at different levels of government and society. The figure above, adapted from Bonebrake et al.,[57] shows what this may look like.

With this approach, it would appear rather unlikely that international environmental agreements would start to see important member countries leave. This would require, however, that the focus on science and relevant other criteria prevail over emotional or political interests.[58] Otherwise, effective biodiversity conservation, even within a framework of TRC, cannot be possible.

Conclusion

It is easy to announce that a party might withdraw from a regime. Especially when this announcement comes at a time when elections are imminent. But as the above and many other chapters in this volume have shown, Japan's withdrawal from the Whaling Convention did not occur for election purposes, but rather over a long-term struggle to get its voice heard within the IWC. This means that just because one state has decided to withdraw from a convention does not necessarily mean that other states in other bodies will follow suit, especially when the conditions are significantly different and it even might benefit from remaining in it.

But as the example of the Bonn Convention, usually having been marked by consensus, has shown, it appears as if the tides are turning to some degree. One can no longer ignore the paradigmatic differences that exist between states, especially with regard to the way charismatic fauna species can or cannot be used. Namibia's and Botswana's announcements—or even threats—furthermore point to the fact that the current *modus operandi* within CITES does not take into account the interests of states, the views of which deviate from majority opinion. Instead of a spill-over, the Japanese case and the case of the SADC countries draw attention to the fundamental flaws some regimes display, essentially challenging the majority voting system and the decision-making processes that underly environmental governance. In the case of CITES, a structural reform appears necessary in order to focus on the convention's purpose and thereby sharpen the convention's efficacy.

One way to achieve this—at least in theory—could be qualified majority voting system that orients itself along the lines of the system within the Council of the European Union. In this way, those states that actually host a species could be given a higher weight *vis-à-vis* states that are not range states of a species. On a global scale, TRC could relegate measures to tackle specific threats to bodies and organisations that are capable of doing so. This, of course, would require a fundamental change in international environmental governance and appears unlikely. But in this time and age, and especially in light of the worrying findings of the IPBES report—in fact, IPBES also proposes transformative change[59]—thinking 'outside of the box' appears more pressing than ever in order to halt the rapidly occurring loss in biodiversity.

Notes

1 Ed Couzens, *Whales and elephants in international conservation law and politics. A comparative study* (Routledge 2013).
2 CITES. Resolution Conf. 9.24 (Rev. CoP17). Criteria for amendment of Appendices I and II, 1994.
3 Justin McCurry, 'Japan to resume commercial whaling one day after leaving the IWC', *The Guardian* (London, 25 January 2019) <https://www.theguardian.com/world/2019/jan/25/japan-to-resume-commercial-whaling-one-day-after-leaving-the-iwc> accessed 24 April 2023.
4 Fergus Hunter and Steve Jacobs, '"Extremely disappointed": Australia lashes Japan's decision to restart commercial whaling' *The Sydney Morning Herald* (Sydney, 26 December 2018) <https://www.smh.com.au/environment/sustainability/japan-to-start-commercial-whaling-in-july-20181226-p50oa8.html> accessed 24 April 2023.
5 Reuters, 'Namibia considers withdrawal from wildlife convention unless rhino trade eased' *Reuters* (London, 19 August 2019) <https://www.reuters.com/article/us-namibia-cites-idUSKCN1VH1WM> accessed 24 April 2023.
6 Scott Barrett, *Environment and statecraft. The strategy of environmental treaty-making* (OUP 2003) 159.
7 Hannah Woolaver, 'From Joining to Leaving: Domestic Law's Role in the International Legal Validity of Treaty Withdrawal' [2019] 30 EJIL 1, 73.
8 Timothy C Bonebrake, Fengyi Guo, Caroline Dingle, David M Baker, Roger L Kitching and Louise A. Ashton, 'Integrating proximal and horizon threats to biodiversity for conservation', 34 *TEE* 9, 781.
9 CITES 'The CITES species' <https://cites.org/eng/disc/species.php> accessed 24 April 2023.
10 Dan WS Challender and Douglas C MacMillan, 'Investigating the influence of non-state actors on amendments to the CITES appendices', 22 *JIWLP* 2, 90, 91.
11 Henriette Kievit, 'Conservation of the Nile crocodile: Has CITES helped or hindered?' in Jon Hutton and Barnabas Dickson (eds.), *Endangered species, threatened convention. The past, present and future of CITES* (Routledge, 2000) 88.

12 Nikolas Sellheim, 'The CITES appendix II-Listing of mako sharks — Revisiting counter arguments', (2020), 115 *Marine Policy*. https://doi.org/10.1016/j.marpol.2020.103887.

13 CITES, Sixty-ninth meeting of the Standing Committee Geneva (Switzerland), 27 November–1 December 2017, Summary Record. <https://cites.org/sites/default/files/eng/com/sc/69/sum/E-SC69-SR.pdf> accessed 24 April 2023.

14 Peter H Sand, 'Japan's 'Research Whaling' in the Antarctic Southern Ocean and the north Pacific Ocean in the face of the endangered species convention (CITES)', (2008), 17 *RECIEL* 1, 56.

15 CITES, *Summary report of the Committee I meeting* (CITES 1994) 185.

16 Sand, 'Japan's Research Whaling' 61 fn 64.

17 Martti Koskenniemi, *Fragmentation of international law: Difficulties arising from the diversification and expansion of international law* (United Nations 2006) <https://legal.un.org/ilc/documentation/english/a_cn4_l682.pdf> accessed 24 April 2023.

18 Leah R Gerber, 'The scientific whaling loophole', (2012), *Science* 337 (6098), 1038; Angus Nurse, 'The beginning of the end? The International Court of Justice's decision on Japanese Antarctic whaling', (2014), *JAWL*, 14.

19 See Sara Wissmann and Maurus Wollensak, 'Sometimes goodbyes are not forever: Japan's hypothetical re-accession to the international convention for the regulation of whaling', (2020), 34 *OYB* 164; see also Couzen's chapter in this volume.

20 Council of the European Union, 'Qualified majority' <https://www.consilium.europa.eu/en/council-eu/voting-system/qualified-majority/> accessed 24 April 2023.

21 Andrew T Guzman, *How international law works. A rational choice theory* (OUP 2007) 121.

22 See, e.g. Steven Bernstein, 'Legitimacy in global environmental governance', (2005), 1 *JILIR* 1–2, 139; Mattei Dogan, 'Political legitimacy: New criteria and anachronistic theories', (2010), 60 *ISSJ*, 195.

23 Helmuth Breitmeier, *The legitimacy of international regimes* (Ashgate 2008) 21.

24 Ibid., 20.

25 See also Malgosia Fitzmaurice, 'The whaling convention and thorny issues of interpretation', in Malgosia Fitzmaurice and Dai Tamada (eds.), *Whaling in the Antarctic. Significance and implications of the ICJ judgment* (Brill 2016) 55.

26 IWC, 'The Way Forward of the IWC. IWC Reform Proposal including a draft Resolution and proposed Schedule Amendment', IWC/67/08. <https://www.jfa.maff.go.jp/j/whale/attach/pdf/index-13.pdf> accessed 24 April 2023.

27 Barrett, *Environment*.

28 NOAA, 'Assessing the Global Climate in 2021' <https://www.ncei.noaa.gov/news/global-climate-202112> accessed 23 April 2023.

29 IPBES, *Global assessment report on biodiversity and ecosystem services of the Intergovernmental Science-Policy Platform on Biodiversity and Ecosystem Services* (IPBES Secretariat 2019). https://doi.org/10.5281/zenodo.3831673.

30 Alister Doyle, 'Iceland and Norway resume whale exports to Japan' *Reuters* (London, 2 June 2008) <https://www.reuters.com/article/us-whaling-idUSL0257551520080602> accessed 24 April 2023.

31 CITES, 'Namibia' <https://cites.org/eng/parties/country-profiles/na> accessed 24 April 2023.

32 CITES, 'Proposals for amendment of Appendices I and II, Eighth meeting of the conference of the parties', (1992) <https://cites.org/eng/cop/08/prop/index.php> accessed 24 April 2023.

33 CITES, 'Amendments to Appendices I and II of the Convention adopted by the conference of the parties at its 10th meeting', (1997) <https://cites.org/sites/default/files/eng/cop/10/E10-amendments.pdf> accessed 24 April 2023.

34 CITES, 'Decision-making mechanism for a process of trade in ivory', CoP17 Doc. 84.3 (2016) <https://cites.org/sites/default/files/eng/cop/17/WorkingDocs/E-CoP17-84-03.pdf> accessed 24 April 2023.

35 CITES, 'Participatory mechanism for rural communities', CoP18 Doc. 17.3 [2019] <https://cites.org/sites/default/files/eng/cop/18/doc/E-CoP18-017-03.pdf> accessed 24 April 2023.
36 Leroy Dzenga, 'It's too early to talk about pulling out of CITES', *The Sunday Mail* (London, 19 June 2022) <https://www.sundaymail.co.zw/its-too-early-to-talk-of-pulling-out-of-cites> accessed 24 April 2023.
37 CITES, 'Proposed Amendments to Resolution Conf. 9.24 (Rev. CoP19)', CoP19 Doc. 87.1 [2022] < https://cites.org/sites/default/files/documents/E-CoP19-87-01_0.pdf> accessed 24 April 2023.
38 The Guardian, 'Namibia election: president wins second term despite scandal and recession' *The Guardian* (London, 1 December 2019) <https://www.theguardian.com/world/2019/dec/01/namibia-election-president-wins-second-term-despite-scandal-and-recession> accessed 24 April 2023.
39 Species+ < https://www.speciesplus.net/species#/taxon_concepts?taxonomy=cites_eu&geo_entities_ids=114&geo_entity_scope=cites&page=1> accessed 23 April 2023.
40 Stasja Koot, 'The limits of economic benefits: Adding social affordances to the analysis of trophy hunting of the Khwe and Ju/'hoansi in Namibian community-based natural resource management', (2019), 32 *SNR* 4, 417.
41 Kenneth Iain MacDonald, 'Global hunting grounds: Power, scale and ecology in the negotiation of conservation', (2005), 12 *CG* 3, 259.
42 Stasja Koot, 'Trophy hunting for conservation and development in Namibia? The limitations of economic benefits and the role of science', *Blog Post* (12 February 2019) <https://stasjakoot.com/2019/02/12/trophy-hunting-for-conservation-and-development-in-namibia-the-limitations-of-economic-benefits-and-the-role-of-science/?fbclid=IwAR1sq4kRzUht5OtRiz6QSHfvcACGRjLkD7GVQGLPouRAmlkN-3wo1bWftcrw> accessed 21 April 2023.
43 Attial Paksi and Adelita Pyhälä, 'Socio-economic impacts of a National Park on local indigenous livelihoods: The case of the Bwabwata National Park in Namibia', (2018), 99 *SES*, 197.
44 IWC, *Annual report of the International Whaling Commission 2006* (IWC Secretariat 2006) 23.
45 CITES, 'Trade with states not party to the convention, Resolution Conf. 9.5 (Rev. CoP16)', (2016) < https://cites.org/sites/default/files/document/E-Res-09-05-R16.pdf> accessed 24 April 2023.
46 CMS, 'Parties and range states' <https://www.cms.int/en/parties-range-states> accessed 14 January 2023.
47 Nikolas Sellheim and Jochen Schumacher, 'Increasing the effectiveness of the Bonn convention on the conservation of migratory species', (2022), 25 *JIWLP* 4, 158.
48 Ibid.
49 Bonebrake et al. 'Integrating'.
50 Evan H. Campbell Grant, David A. W. Miller, Benedikt R. Schmidt, Michael J. Adams, Staci M. Amburgey, Thierry Chambert, Sam S. Cruickshank, Robert N. Fisher, David M. Green, Blake R. Hossack, Pieter T. J. Johnson, Maxwell B. Joseph, Tracy A. G. Rittenhouse, Maureen E. Ryan, J. Hardin Waddle, Susan C. Walls, Larissa L. Bailey, Gary M. Fellers, Thomas A. Gorman, Andrew M. Ray, David S. Pilliod, Steven J. Price, Daniel Saenz, Walt Sadinski and Erin Muths, 'Quantitative evidence for the effects of multiple drivers on continental-scale amphibian declines', (2016), *SR* 6 (25625). https://doi.org/10.1038/srep25625.
51 Elli Louka, *International environmental law. Fairness, effectiveness and world order* (CUP 2006).
52 Fitzmaurice, 'Interpretation'.
53 CITES, 'Proposed Amendments'.

54 Tim O'Riordan, 'Protecting beyond the protected', in Tim O'Riordan and Susanne Stoll-Kleemann (eds.), *Biodiversity, sustainability and human communities. Protecting beyond the protected* (CUP 2002) 61.

55 CITES, 'CITES strategic vision: 2021–2030', Resolution Conf. 18.3. < https://cites.org/sites/default/files/documents/COP/19/resolution/E-Res-18-03.pdf> accessed 23 April 2023.

56 Tanya Wyatt, *Is CITES protecting wildlife? Assessing implementation and compliance* (Routledge 2021).

57 Bonebrake et al., 'Integrating', 782.

58 See also Nikolas Sellheim and Sheryl Fink, 'The role and impact of NGOs in marine mammal governance', in Nikolas Sellheim and Dwayne R Menezes (eds.), *Non-state actors in the Arctic region* (Springer 2022) 231.

59 IPBES, 'Transformative change assessment' < https://www.ipbes.net/transformative-change> accessed 23 April 2023.

Part II
Cultural Considerations

7   Indigenous Whaling After Japan's Withdrawal from the International Whaling Commission

Malgosia Fitzmaurice and Agnes Rydberg

Introduction

On 26 December 2018, Japan announced that it would withdraw from the IWC and expressed its intention to begin commercial whaling within its coastal waters and Exclusive Economic Zone (EEZ) for the first time in 30 years.[1] As of 30 June 2019, Japan is no longer bound by the International Convention for the Regulation of Whaling (ICRW).[2] The ICRW distinguishes between three types of whaling activities: commercial whaling, scientific research whaling, and aboriginal subsistence whaling (ASW). Commercial whaling has basically been terminated by the IWC's moratorium on commercial whaling, which in 1982 established a 'zero-catch-limit'. The moratorium is binding on all Member States except for Norway, Russia, and Iceland.[3] Russia however stopped commercial whaling in 1990, and in 2020. Accordingly, only Norway and Iceland are at present engaged in commercial whaling in their EEZs. Scientific whaling is allowed by Article VIII of the ICRW.[4] The special scientific permits are granted by the Member States but under review and monitoring by the IWC. Scientific whaling conducted by Japan culminated in the case before the ICJ in 2014, which concluded that whaling under the Second Phase of Japan's Whale Research Program under Special Permit in the Antarctic at that time – JARPAII – activities were not for the purpose of scientific research and therefore fell outside the ambit of the ICRW.[5]

The third type of whaling activity, ASW, is not expressly recognised in the ICRW itself and lacks clear legal contours. It is referred to in the Schedule to the ICRW and managed by a subcommittee of the IWC. Today, four member countries conduct ASW: Denmark (Greenland), Russia (Chukotka), St Vincent and the Grenadines (Bequia), and the United States (Alaska, and also potentially a resumption of hunts previously undertaken by the Makah Tribe of Washington State).[6] In addition, ASW is undertaken by nationals of certain non-Member States outside the purview of the ICRW, including Canada, Equatorial Guinea, Indonesia, and the Philippines.[7]

ASW is subject to deep intercultural controversy, particularly over the appropriateness of killing whales as such, and over the aboriginal killing methods, which may be more cruel than modern and quicker techniques.[8] It is, in its core, protected by Article 27 of the International Covenant on Civil and Political Rights (ICCPR)[9]

DOI: 10.4324/9781003250814-9

and Article 25 of the United Nations (UN) Declaration on the Rights of Indigenous Peoples (UNDRIP).[10] However, relevant legal terms and key concepts, such as 'non-commercial hunt', 'indigenous people', and 'indigenous culture' are difficult to apply, and it may therefore be difficult to distinguish ASW from commercial whaling. In addition, the type of whaling foreseen in the ICRW has since 1949 been substantially modified through IWC practice and case-law, with commercial and scientific whaling being basically zero.

This chapter provides an overview of indigenous whaling and analyses whether Japan's withdrawal had an impact on the regulation of ASW within the IWC. It argues that in compensation for the closure of commercial and scientific whaling, ASW will assume the leading role in types of whaling regulated by the IWC, and that the withdrawal of Japan has turned the attention of the international community of IWC States, non-governmental organisations (NGOs), and civil society to indigenous whaling.[11] It also demonstrates that the issue of ASW remains controversial, particularly due to: (i) the imprecise and ambiguous definition of ASW, rendering the distinction between indigenous and commercial whaling blurred; (ii) the recent trend to increase the quotas assigned to indigenous whaling by the IWC; (iii) the inefficient and cruel aboriginal killing methods compared to modern killing methods; and (iv) the question of whether the resumption of whaling by Makah people qualifies as ASW or not.

As ASW should be analysed within the general context of rights of indigenous peoples, this chapter proceeds in Section 'Indigenous Peoples' Rights' to provide a brief assessment of the legal regime pertaining to indigenous peoples' cultural diversity. Section 'The IWC and Indigenous Whaling' gives an historical outline of ASW within the IWC, and Section 'Current Regulation of ASW under the IWC' discusses ASW under the IWC at present. Section 'Aboriginal Whaling under the IWC Purview' analyses ongoing ASW operations under the IWC purview and explores the potential impact of Japanese IWC-withdrawal on the focus of the ICW and ASW more generally. Section 'Concluding Remarks' concludes.

Indigenous Peoples' Rights

A full legal analysis of the legal regime pertaining to indigenous peoples' cultural diversity is beyond the scope of this chapter, but since some indigenous peoples consider whaling as an expression of their cultural identity, a general introduction is called for. Indigenous peoples may be defined as 'aboriginal, indigenous or native peoples who share strong community, familial, social and cultural ties related to a continuing traditional dependence on whaling and on the use of whales'.[12] Article 27 of the ICCPR, which is central to the rights of minorities and indigenous peoples,[13] stipulates that:

> In those States in which ethnic, religious or linguistic minorities exist, persons belonging to such minorities shall not be denied the right, in community with the other members of their group, to enjoy their own culture, to profess and practise their own religion, or to use their own language.

It may be noted that the wording of Article 27 of the ICCPR may appear also to apply to Japanese coastal whaling communities. However, despite several attempts of Japan, the legal status of Indigenous whaling was not accorded to these communities.[14]

Furthermore, General Comment 21 to the Covenant on Economic, Social, and Cultural Rights refers to the right to culture of indigenous peoples.[15] Paragraph 37 is of particular importance, stipulating that indigenous peoples enjoy collective rights:

> Indigenous peoples have the right to act collectively to ensure respect for their right to maintain, control, protect and develop their cultural heritage, traditional knowledge and traditional cultural expressions, as well as the manifestations of their sciences, technologies and cultures, including human and genetic resources [...]. States parties should respect the principle of free, prior and informed consent of indigenous peoples in all matters covered by their specific rights.

It should, however, be mentioned that the character of indigenous peoples' collective rights (including cultural rights) is somewhat controversial. The legal integrity of collective rights of indigenous peoples, as opposed to their individual rights, is not universally accepted. Nevertheless, the 2007 UNDRIP is unambiguous and stipulates that '[i]ndigenous people possess collective rights which are indispensable for their existence, well-being and integral development as peoples'.[16] Such a collective right is vital for the self-realisation and survival of indigenous peoples as a societal group.[17]

The right to culture based on dignity is also a cornerstone of the approach adopted in relation to indigenous peoples, as enshrined in Article 15(1) of UNDRIP. This provision provides that '[i]ndigenous peoples have the right to the dignity and diversity of their cultures, traditions, histories and aspirations which shall be appropriately reflected in education and public information'. The cultural rights of indigenous peoples have to be approached and understood in harmony with the fundamental rights underlying the Declaration, as part of the holistic nexus of indigenous rights. Therefore, such rights have to be viewed together with the right of self-determination,[18] and the right to autonomy or self-government in matters relating to their internal and local affairs and to financing their autonomous function.[19]

Furthermore, indigenous peoples' right to culture is intertwined with their rights to land and natural resources (Articles 25 and 26), and their special relationship with land and nature distinguishes them from other types of social minorities.[20] The cultural rights of indigenous peoples are also set out in UNDRIP Articles 11–14. Article 11 grants indigenous peoples the right to practise and revitalise their traditions and culture, Article 12 concerns indigenous peoples' right to practise their spiritual and religious traditions, Article 13 refers to indigenous peoples' intangible heritage, and Article 14 concerns their right to establish and control their educational systems. The Human Rights Committee (HRC) has also reaffirmed the right to the culture of indigenous peoples in a range of cases. In *Kitok v. Sweden*[21] and *Ominayak v. Canada*,[22] the Commission extended the cultural guarantees under

Article 27 ICCPR to the economic and social activities relied upon by the indigenous group.[23] It is, however, important to note that such a right of indigenous peoples to cultural diversity is neither unlimited nor absolute; rather, it is subject to a balancing act by the State authorities, as is the case of other human rights, for instance the right to privacy. Therefore, as the HRC observed in *Länsman and Others v. Finland*, the interests of the broader society must be taken into account.[24]

Accordingly, from the general state of human rights law and the practice of human rights bodies, it seems that indigenous peoples enjoy the right to whaling so long as it is a *bona fide* expression of their traditional way of life and their traditional use of natural resources. However, it has been recognised, in the case of indigenous peoples, that there has been a trend towards a greater recognition of customary law and self-governing rights, without any retreat or backlash.[25] Nonetheless, certain indigenous practices have been viewed as culturally conservative and restricting upon individual freedoms due to their overly communitarian and/or traditionalist character, and as expressing a desire for cultural isolationism.[26] Some argue that, inadvertently, social conservatism, traditionalism, and isolationism are elevated to 'sacred obligations' in order to silence, manage, or otherwise 'delegitimise' group members who want to change such practices.[27] After all, Indigenous groups are not immune from inevitable social changes within their own communities, and a regime of collective rights – while unquestionably progressive given that it safeguards the cultural identity and relative autonomy of groups surrounded by outsider dominant cultures – should not lend itself to intra-community oppression when its principal overall purpose is to resist domination and oppression from forces extraneous to the group, such as the State and the dominant culture that envelops such ethnic groups.

The IWC and Indigenous Whaling

International recognition of ASW predated the ICRW; the special position of whaling rights of aboriginal communities was already recognised and included in the 1931 Geneva Whaling Convention.[28] Article 3 of this Convention stated that it did not apply to aborigines dwelling on the coasts of the Parties, provided that:

1 they only use canoes, pirogues, or other exclusively native craft propelled by oars or sails[29];
2 they do not carry firearms;
3 they are not in the employment of persons other than aborigines;
4 they are not under contract to deliver the products of their whaling to any third person.

However, an equivalent provision was neither included in the 1937 London Agreement nor in the 1938 Protocol amending the 1937 London Agreement.[30] While the 1946 ICRW does not itself include any special provision regulating ASW, the Schedule to the ICRW recognises its special position by excluding it from the definition of, and the provisions relating to, commercial whaling.

When the ICRW was originally concluded, the second paragraph of the Schedule prohibited the taking and killing of gray and right whales due to their fragile status. It stated that '[i]t is forbidden to take or kill gray whales or right whales, except when the meat and products of such whales are to be used exclusively for local consumption by the aborigines'.[31] This prohibition did not apply where the whaling of such species was conducted for local consumption of meat and other products by indigenous peoples. Thus, it appears that even when conducted by indigenous whalers, the whaling of gray or right whales either for commercial purposes or when the meat or products were to be distributed extensively outside the communities of those indigenous peoples was prohibited.[32]

In 1979, the IWC Anthropology Panel adopted an unofficial definition of 'subsistence whaling' as comprising:

1 the personal consumption of whale products for food, fuel, shelter, clothing, tools, or transportation by participants in the whale harvest;
2 the barter, trade, or sharing of whale products in their harvested form with relatives of the participants in the harvest, with others in the local community or with persons in locations other than the local community with whom local residents share familial, social, cultural, or economic ties. A generalised currency is involved in this barter and trade, but the predominant portion of the products from each whale is ordinarily directly consumed or utilised in their harvested form within the local community; and
3 the making and selling of handicraft articles from whale products, when the whale is harvested for the purposes of (1) and (2) above.

ASW was again defined in 1981 by the IWC's Technical Committee Working Group on Development of Management Principles and Guidelines for Subsistence Catches of Whales by Indigenous (Aboriginal) Peoples as comprising whaling conducted for

> purposes of local aboriginal consumption, carried out by or on behalf of aboriginal, indigenous or native people who share strong community, familial, social and cultural ties related to a continuing traditional dependence on whaling and on the use of whales.[33]

These definitions are capable of giving rise to complex issues. For instance, the interchangeable use of the terms 'aboriginal', 'native', and 'indigenous' is confusing, particularly as within different indigenous communities, these terms have different meanings.[34] This ambiguity perhaps could permit the inclusion of Japanese, Icelandic, and Norwegian whaling, which to a great extent is argued as culture-based.

There has for example been discussion as to whether whaling in Greenland can qualify as aboriginal. The 1979 definition of 'subsistence use of whale products', coined by cultural anthropologists, is more restrictive as to the area in which the distribution of whale products is permitted; and it does not recognise the distribution of whale products that involve monetary exchanges, as in ASW.[35] It may also be noted that the report proposing the definitions was perhaps inconsistent as it states that, in some cases, products are distributed to, and used by, communities away from the coastal areas where whaling is actually conducted; and in some areas, the practice of trading to meet subsistence needs has emerged. Further, the IWC ad hoc Working Group stated that it was arguable whether there is a difference in principle between the sale of whale products in order to buy essential goods and the direct exchange of whale products for such goods.

According to Hamaguchi, this is indicative of the fact that even the ad hoc Working Group's definition did not completely deny all cases of extensive distribution of whale products or distribution involving monetary exchanges for whale-related goods.[36] At the outset, therefore, it appears that confusion arose due to the lack of any conclusive definition of what constitutes 'commercial' whaling, which, at times, makes the distinction between 'aboriginal' and 'commercial' whaling blurred, despite the efforts of the ad hoc Working Group to distinguish between them. These two forms of whaling are considered to be different due to their different objectives, which also include management and catching. The main objective of the management of ASW was to maintain individual stocks at the highest possible level, and the main purpose of catching whales was to fulfil the nutritional and cultural needs of the indigenous peoples contemplated by the relevant agreements. The main objective of commercial whaling was to maximise yields from individual stocks and to ensure the longevity of the respective whaling industries of the State Parties to the ICRW, and the main purpose of catching whales was to sell their products. To this end, it has been observed that 'these differences indicate that ASW prioritises quality (the cultural aspect) and commercial whaling prioritises quantity (the economic aspect)'.[37]

Nevertheless, the distinction between commercial whaling and ASW remains ambiguous.[38] The ASW Sub-Committee (ASWS) was established in order to consider documentation on needs relating to aboriginal whaling and to advise the IWC Technical Committee on the setting up of proper management measures. The field of ASW was, consequently, subdivided into the following: (1) subsistence whaling; (2) nutritional whaling; and (3) cultural whaling. As it was above-mentioned, the definition of Indigenous (Aboriginal) Whaling is rather obscure, without well set contours and rather broad, including such elements as culture and nutrition. Nations, such as Norway and Iceland, with a very long tradition of (commercial) whaling have argued that it forms an element of their culture and history. Whaling obviously also has a very strong nutritional aspect in these States. Therefore, a suggestion may be made that, taking all these into account, ASW may be also extended to modern whaling nations with a long-standing tradition.

Thus, the definitions of aboriginal set out above indicate that aboriginal whaling should be local and non-commercial in nature. However, this conflicts to some extent with the definition of 'consumption', which allows the sale of by-products.[39] Therefore, 'in practice [...] the "subsistence requirement" seems to be a kind of main category, including "food" by definition, whereas nutritional and cultural needs are subcategories to "Aboriginal Subsistence Whaling", which are closely connected to each other'.[40] Apart from this, however, there are a number of problems relating to the definition of the three categories themselves and to their inter-relationship. Indeed, State practice indicates that the submission of the required evidence proving 'nutritional and cultural need', which is a fundamental requirement for allocation of quotas for aboriginal peoples, has frequently been very difficult to achieve.[41]

Furthermore, the distinction sought to be made by the IWC between subsistence and nutritional needs conflicts with the interpretation of the UNHRC, which interpreted traditional livelihoods and means of subsistence (including traditional diet) as an integral part of indigenous culture under Article 27 ICCPR.[42] Additionally, the UNHRC allows the inclusion of some commercial elements within the definition of 'aboriginal subsistence' and, indeed, stresses the economic viability of a livelihood as a material criterion of the fulfilment of the provisions of Article 27 ICCPR.[43] Although Indigenous peoples are heard at the forum of the IWC through their states, significant changes have taken place in relation to their position therein. It was recommended at the workshop held in 2015 that the IWC:

> [C]onsider exploring options regarding how the IWC could stay better informed of current developments regarding indigenous peoples' rights; look at mechanisms to improve the status of indigenous delegates to IWC meetings to establish a more timely, distinct and steady approach to ASW issues; discuss the appointment of an IWC representative to attend a session of the UN Permanent Forum on Indigenous Issues; and explore the potential benefits of joining the UN Inter-Agency Support Group on Indigenous Issues, through its Secretariat.

The workshop also recommended that the term 'need statement' be replaced by the term 'Description on the [insert name] hunt relevant to catch/strike limit requests' and that a draft outline be developed, taking into account the need for flexibility, to avoid any indication of prescription or compulsion, and to minimise the effort involved and avoid duplication.[44]

Current Regulation of ASW under the IWC

Today, the IWC's three objectives for management of ASW are as follows:

1 To ensure that the risks of extinction to individual stocks are not seriously increased by subsistence whaling;
2 To enable aboriginal people to harvest whales in perpetuity at levels appropriate to their cultural and nutritional requirements, subject to the other objectives;
3 To maintain the status of whale stocks at or above the level giving the highest net recruitment and to ensure that stocks below that level are moved towards it so far as the environment permits.[45]

In general, the IWC has identified four specific whaling operations as qualifying for the status of ASW:

i minke and fin whales (formerly also humpback whales) in Greenland;
ii humpback whales in the Lesser Antilles (specifically the island of Bequia, St Vincent and the Grenadines);
iii bowhead whales (and formerly also gray whales) in the US (Alaska), and gray whales in Russia (Chukotka); and
iv bowhead and gray whales in the US (Alaska).[46]

On the basis of the above objectives and criteria, ASW is allowed at present for the following countries:

i Denmark (in relation to the Inuit peoples of Greenland) – fin and minke whales;
ii Russian Federation (Siberia) – gray and bowhead whales;
iii St Vincent and the Grenadines (Bequia) – humpback whales; and
iv the US (Alaska and Makah Indigenous peoples) – bowhead and gray whales.[47]

In brief, there are two fundamental requirements for the allocation of quotas for ASW: ensuring that hunts do not seriously increase risks of extinction and that hunted whale populations are then maintained at healthy, relatively high levels; and enabling native people to hunt whales at levels appropriate to cultural and nutritional requirements (known as 'cultural and nutritional needs').

National governments provide the IWC with evidence of the needs of their indigenous people. This is presented in the form of a 'Needs Statement' that details the cultural, subsistence, and nutritional aspects of the hunt, products, and distribution. The Scientific Committee provides advice on the sustainability of proposed hunts and safe catch limits.[48]

An ASWS was established in 1983 to review quota applications and provide advice on technical management measures. The IWC has subsequently focused on reducing the numbers of whales struck but not landed,[49] ensuring the sustainability of specific aboriginal hunts[50] and improving humane killing methods.[51] At the IWC meeting in September 2018, a number of new initiatives were endorsed in order to facilitate a more straight-forward process when catch limits are next considered at the 2024 meeting of the Commission. The new initiatives include a:

a new timetable for sharing information from the hunts, and receiving feedback, maximising discussion time and transparency;
b agreement that status quo catch limits would be renewed automatically, assuming a series of agreed steps continue to be completed;
c commitment to establish closer ties with international and inter-governmental organisations focusing on indigenous rights.[52]

Furthermore, the IWC has for many years been developing the Aboriginal Whaling Management Scheme (AWMS), which aims to include two elements:

i a quota-setting mechanism (which is already in place); and
ii a supervision and control scheme in order to establish the future management of ASW.

Until 2018, Aboriginal subsistence quotas under the quota-setting mechanism were adopted for a period of five years. Since 2018, the period was extended to six years and automatically renewed if no IWC member objects.[53] Under the IWC, there is a marked limitation in the number of species of whales that have been singled out as eligible for Aboriginal hunting, as indicated by the inclusion of bowhead whales in the Moratorium. Eventually, the AWMS (including the Aboriginal Whaling Management Procedure once it becomes fully operational) will provide guidelines and will set requirements for surveys and data as well as case-specific elements. It is also likely to cover certain scientific, logistical, and regulatory aspects of aboriginal whaling, including the inspection/observation of catches.

Discussions regarding Aboriginal whaling quotas can be highly divisive in the IWC. An example of this is the 2009 meeting, where the primary issue was a discussion regarding a request for a catch of ten humpback whales for Greenland.[54] The debate focused on the statement from the IWC Scientific Committee as to whether such a catch would not harm the stock of these whales and whether Denmark (on behalf of Greenland) had adequately made the case for the 'needs' of Inuit people for humpback whales, both of which are necessary conditions for the allocation of quotas for ASW. In the event, Denmark's request was unsuccessful, although it gained approval later on.[55] It may be added that there are also voices that support the view that there are insufficient quotas assigned to Russia's Chukchi people. It accordingly appears to be a general trend in relation to Indigenous peoples that they will try to argue for higher allocations than those awarded to them.[56]

That said, there are many limitations concerning Indigenous (Aboriginal) whaling. They relate to the species that can be hunted, the method of hunting, and an overall duty of ecological approach in general environmental law of preservation of endangered species, which is based on the premise that the taking of animals by Indigenous peoples can only be done if sustainable.[57] This is consistent with a Resolution adopted by the IWC in 1983 on Indigenous whaling, which states that ASW should be 'consistent with effective conservation of whale stocks'.[58] The term 'effective conservation' is very vague and it is unclear to which aspect of conservation it refers.[59] In addition, Indigenous peoples have little say in the matters of Aboriginal whaling; it is up to respective governments to represent them in the IWC.[60] This lack of legal clarity and consistency in the drafting of paragraph 13 and in state practice makes the politicisation of Aboriginal whaling inevitable. By gaining a three-quarters majority in the IWC, any form of whaling that possesses some elements of ASW could potentially be considered and thus endorsed by the

IWC as ASW. Consequently, such whaling, which seems to have all the hallmarks of ASW, would not be endorsed by the IWC if, say, marginally more than a quarter of IWC members vote against it.

It has been suggested that ASW should be approved only on condition that there is a cultural, nutritional, and economic need for it, and that the whales being harvested are not threatened with extinction.[61] Presumably, if the IWC members feel strongly about this lack of legal certainty and can form some majority to push for this, they could adopt some statement or declaration to shed more light on what constitutes ICRW-compliant ASW. Such a development could address the seeming inconsistency in the IWC and thus enhance the integrity of the regime in the minds of all its members. It is reasonable to assume that Aboriginal whaling, whilst not without its controversies, seems more acceptable for many than commercial whaling, given that the former involves strong arguments in relation to the cultural rights of marginalised social groups. However, the fact that Aboriginal whaling may be less opposed than commercial whaling does not mean that it is not contentious.

A detailed analysis of ASW leads to the conclusion that it is equally problematic and involves its fair share of contentious issues that are questions of law and issues of ethics. The lack of agreement as to the definitive content of the term ASW and the ongoing disputes regarding the number of whale stocks open to Aboriginal whaling result in the inability of the IWC to manage and regulate ASW effectively and consistently. This is further compounded by the fact that ASW also takes place outside the remit of the IWC and, with regard to such whaling, the real dearth, in many cases, of statistical data, along with the unilateral regulation on the part of other states, significantly undermines the chances for the effective multilateral regulation of this type of whaling.

Aboriginal Whaling under the IWC Purview

Current Operations

A considerable amount of ASW takes place outside the purview of the IWC in States that are not party to the ICRW, such as Equatorial Guinea, Indonesia, Canada, and the Philippines. For these States, there are no quotas allocated by the IWC for aboriginal whaling. These examples illustrate the difficulties in assessing the impact of aboriginal whaling on the whale population, due to the lack of correct data is some countries.[62] Within the IWC, whaling by the Makah people, an indigenous people of the Pacific Northwest Coast living in Washington State is one of the most controversial cases of ASW. It has been the subject of much discussion at the IWC and still remains controversial in various ways, including the following:

i it is an instance of a claim (there are others) to resume Aboriginal whaling after a period during which, for different reasons, whaling had been abandoned, in this case, a claim by the Makah people to resume whaling after a hiatus of several decades;

ii there were also doubts as to whether their whaling was purely ASW or whether a commercial dimension was also present;

iii the claim raised ethical concerns regarding the resumption of Aboriginal whaling; and

iv it also raised the issue as to whether Aboriginal whaling constitutes a cultural exemption.[63]

These outstanding questions in relation to the Makah people can be said to reflect general confusion surrounding Aboriginal whale hunting. The last legal whale hunt the tribe was able to perform was in 1999, when it held its first hunt in more than 70 years. Nevertheless, in September 2021, an administrative law judge issued a recommendation to the US Department of Commerce, arguing the tribe should be granted a waiver under the Marine Mammal Protection Act to resume whaling.[64] The waiver would allow Makah tribe members to land up to three Eastern North Pacific gray whales in an even year and one in an odd year over a 10-year period. A final decision rests with an administrator in the National Marine Fisheries Service, a branch of the Commerce Department, and the question remains whether the Tribe's whaling activities would qualify as ASW or not.[65]

Furthermore, the ASW quotas for the catch of the Bering-Chukchi-Beaufort Seas stock of bowhead whales, taken by native peoples of Alaska and Chukotka, amount to a total of up to 392 bowhead whales in the period 2019–2025, with no more than 67 whales struck in any year, except that any unused portion of a strike quota from the three prior quota blocks shall be carried forward and added to the strike quotas of subsequent years, provided that no more than 50% of the annual strike limit is added to the strike quota for any one year. The quotas for Eastern North Pacific gray whales, taken by native peoples of Chukotka and Washington State, amount to a total of up to 980 gray whales in the period 2019–2025 provided that the number of whales struck in any year shall not exceed 140.

In the context of Greenland, the Inuit people have about 200 years of history of whaling. In contemporary Inuit society, whaling is very important culturally, socially, spiritually, politically, and nutritionally. Whaling is one of the factors defining their community identity. It appears that whaling is absolutely fundamental for Inuit cultural and social continuation of them as a people, given that members of this society regard it as a defining aspect of their cultural identity.[66] In 1999, Greenland began to draft its own regulations on the protection of beluga and narwhal whales. Hunts were to establish annual and regional quotas, prohibit the killing of females and calves, prohibit hunting using nets, and set trading quotas for meat and blubber (though no mention was made of narwhal tusks, which are the most sought-after part of the narwhal whale). However, up to 2004, the catches of both types of whale were not regulated by any legal act.

In 2004, Greenland, under its system of self-government, adopted an Executive Order regarding quotas.[67] In June 2004, local quotas were set as from 2005. No hunting quotas were set for East Greenland, which is the part of the island where professional leisure hunters may hunt for narwhals. However, the hunting quotas for 2004 were three times higher than that recommended by marine biologists. The same applies to the hunting of beluga whales. For example, the estimated catch for the 1998 and 1999 harvests was to reach 700 a year; but in 1998 the catch actually reached 744. As has been noted,

> [t]hese large annual harvests, combined with the alarming scientific population estimates, have also caused the Greenlandic and Canadian Joint Commission on the Conservation and Management of Narwhal and Beluga to define the Greenlandic hunt as non-sustainable, and the Joint Commission has urged the Home Rule [i.e., Greenland's] government to intervene immediately.[68]

Greenland's government attempted to intervene and regulate this hunt; but due to the cultural and economic importance of beluga, this was not completely successful. However, one of the reasons given by Greenland to hunt larger than recommended quotas was the right of the Inuit peoples to cultural diversity. In relation to whaling by Greenland's Inuit within the remit of quotas set by the IWC, in 2012 Denmark requested on behalf of the Inuit an increase in whale quotas. This request was met with opposition from certain IWC members, causing additional tensions. One of the reasons for the rejection of this request was because some of the IWC's members were of the view that the proposed quota included an element of commercialisation and was higher than just to meet subsistence needs.[69]

Furthermore, Greenland's set quotas for the season 2013–2014 (198 whales) were much to the concern of other IWC members. This action of Greenland was treated as an infraction by some States, which made a statement to the Infractions Sub-Committee. Denmark opposed a qualification of Greenland's action as an 'infraction'. This dispute was not resolved, partly due to the lack of a clear definition of what constitutes an infraction.[70] This illustrates the conflicting nature of aboriginal whale hunting and cultural diversity issues; Aboriginal hunts of marine mammals are a highly complex ethical issue. It is true that the Inuit and their cousins have traditionally thrived on what they call natural food, caribou, seal, beluga, whales, and other marine mammals. They do not have much money with which to draw food from the cash economy, and they do not fare well on the kind of food eaten by Europeans and Americans. But it turns out that Greenland's hunt for whales is as much about profits as it is about aboriginal rights, and Greenlanders are not observing the terms of the IWC quota that permits the hunt '[to] be conducted solely for aboriginal subsistence purposes'.[71] The IWC quotas for 2019–2025 amount to the following: (i) an annual strike limit of 20 minke whales; (ii) annual strike limit of 2 West Greenland bowhead whales; (iii) an annual strike limit of 19 West Greenland fin whales; (iv) an annual strike limit of 164 West Greenland common minke whales; and (v) an annual strike limit of 10 West Greenland humpback whales.

ASW Post-Japanese Withdrawal: Outstanding Controversies

The withdrawal of Japan from the ICRW is perhaps very indicative of how better understating of the position of Japan could in the future form Japan's whaling policy. The way forward post-IWC is to emphasise and publicise at the world for an importance of Japanese coastal whaling and indicate its importance for culture, in line with the ASW. The common knowledge of Japan as a whaling hunting country can be shifted by evidencing of whaling to its culture and religion. Historically, whales were considered to be embodied deities (*shintai*), and whale religions sprang up in coastal villages, called Hyochakushin (Drifting Ashore God) or *Yorikami Shinkyo* (The Religion of the Visiting Kami).[72] Hunting of small-type cetaceans has been practised in Japan for many centuries, but the origins of what is referred to as Small-Type Coastal Whaling can be found in the beginning of minke whaling off the Japan coasts in the 1930s, which is characterised, firstly, by the species of whale caught (minke, Baird's beaked and pilot whales), and secondly, by the small size of the whaling vessel (between 15 and 50 tons).[73]

Historically, and prior to Japanese withdrawal, the focus of the IWC has to a large extent centred on scientific whaling. However, the IWC has recently stressed that scientific whaling should be non-lethal.[74] At the IWC meeting in 2018,[75] States Parties adopted the Florianópolis Declaration, which stipulates that the IWC objective is 'non-lethal management', 'maintenance of healthy cetacean populations', and 'the recovery of cetacean populations to their pre-industrial levels'. By the Florianópolis Declaration, the IWC also seeks to 'ensure that aboriginal subsistence whaling for the benefit of indigenous communities should meet the Commission's management and conservation objectives, taking into account the safety of hunters and the welfare of cetaceans'. As commercial whaling is generally in decline – Norway and Iceland being the only States currently conducting this form of whaling within the IWC –[76] and scientific whaling limited to non-lethal activities, the withdrawal of Japan inevitably entails that focus will, for all practical purposes, be on ASW, which will be the only remaining whaling activity.[77] As a matter of fact, the withdrawal of Japan turned the attention of the international community of IWC States, NGOs, and civil society to indigenous whaling, and the focus of the IWC has shifted from scientific to indigenous whaling. We would like to emphasise again that the practices of Japan concerning whaling – being of cultural and nutritional importance (especially small coastal whaling) – could be included in a general paradigm of the ASW to align Japanese whaling with that of rest of this type of whaling, even external to IWC.

That said, ASW is more often than not a thorny question, and certain outstanding issues remain, including: (i) the definition of ASW and the divergence between indigenous and commercial whaling, especially given the fact that the application of the Aboriginal exception to the Moratorium has been inconsistent[78]; (ii) the recent trend by the IWC to increase quotas for ASW catches; (iii) the inhumane and inefficient methods of hunting and killing used by indigenous peoples; (iv) whether the precautionary principle and the obligation to conduct an environmental impact assessment is applied (or should be applied) in the allocation of quotas and hunting operations; and (v) whether the whaling resumption by the Makah qualifies as ASW or not.

The trend to increase ASW quotas and the plea to apply a precautionary approach in this regard has been visible for a number of years and has caused some controversies. For example, in 2002, when St Vincent and the Grenadines submitted at the IWC a 'Needs Statement' for the increased quota of whales, New Zealand, Australia, and the United Kingdom advocated that a precautionary approach be adopted, given the uncertainty of the scientific evidence of the stock, and thus the request should be turned down.[79] Whereas at IWC63 in 2011 saw no changes to the limits regarding ASW, delegates at IWC64 in 2012 approved increased quotas for several aboriginal subsistence hunts, except Greenland's. IWC65, held in 2014, adopted increased four-year catch limits for Greenland ASW. Denmark's proposal to raise Greenland's quotas was supported by the EU and the US,[80] but met strong opposition within the IWC. Many states such as Monaco questioned the methods used to calculate 'need'. The Latin American group of countries questioned Denmark over not reporting Greenland's catches taken since the last meeting as infractions.[81] However, IWC67 likewise increased the annual strike limit for common minke whales off East Greenland to 20 in order to satisfy ASW need in that area and increased the annual strike limit for Eastern North Pacific gray whales to 140 in order to satisfy ASW need.[82]

Furthermore, the 'Need Statement' of Greenland has steadily expanded, raising concerns over an increasingly blurred boundary between subsistence and commercial whaling.[83] This blur can also be manifested through the parallel between ASW and national/nutritional whaling by counties like Japan that is more populous and more industrial but share the same strong tradition of local, coastal whaling as an element of their culture and history.

This has provoked an intriguing internal battle within the cohort of EU Member States party to the ICRW, where Denmark (as the representative of Greenland) has sought to reconcile its obligations towards its Indigenous communities with the EU's common position against excessive ASW quotas.[84] The formidable negotiating presence of the EU bloc has meant that Greenlandic quotas have often been arranged more through collective discussions at Brussels than within the IWC, with the EU proving to be both a vehicle for, and an obstacle against, ASW claims.[85]

The increased quota allocation has also met fierce opposition by NGOs and civil society. The Whale and Dolphin Conservation Society, an NGO, has challenged the tonnage of whale meat that Greenland seeks because it is significantly more than all the marine protein currently being consumed by Greenlandic people from whales, seals, and small cetaceans combined. Iceland and Norway were strongly criticised at the same meeting for their continuation of commercial whaling. The Animal Welfare Institute, opposing the Makah's recent ASW permit, argued that any whaling by the Makah Tribe, however limited, would place endangered groups of gray whales at risk of being harpooned. Moreover, the group argues that no whales should be killed, as the large marine mammals face numerous threats, including pollution, and ship strikes.[86]

The issue of the inhumane and inefficient hunting and killing methods used by indigenous peoples compared to more modern techniques likewise remains controversial. The IWC recognises that traditional methods used to hunt and kill whales for ASW tend to be less efficient than those used in commercial whaling operations, with the result that: (a) times to death are longer; (b) instantaneous death rates are lower; and (c) 'struck and lost' rates are higher. Although the IWC has adopted several resolutions seeking improvements in the humaneness of ASW operations, IWC resolutions are not binding on parties. To date, efforts to improve the welfare of ASW hunts have been ad hoc and undertaken by interested governments rather than through an organised effort by the IWC. Despite the availability of modern weaponry and training of indigenous hunters by whalers from the US and Norway, outdated equipment and methods are still used.[87] Although several civil society organisations have expressed concern regarding methods of Indigenous hunting,[88] modern methods such as explosive harpoon gun used by Iñupiat peoples are more humane.

Thus, ASW is subject to divisive approaches, and in relation to this type of whaling, there is a persistent question of discontent regarding the lack of any uniform approach to what constitutes 'needs'. There are also continuing discussions regarding the dividing line between commercial and Aboriginal whaling, certain views arguing that the IWC has a 'money fetish', which results in an approach that all whaling that brings in money must be by nature unsustainable.[89] That said, the regulation of ASW is a sensitive and political issue, and it remains to be seen what future approaches will be adopted by the IWC in this regard.

Concluding Remarks

At present, some of whales are endangered and some critically endangered.[90] The overexploitation of marine mammals has been on the world agenda for a very long time.[91] At IWC67,[92] States defined the role of the IWC as ensuring 'the recovery of cetacean populations to their pre-industrial levels and in [that] context reaffirms the importance in maintaining the moratorium on commercial whaling'[93] and confirmed that the temporary moratorium on commercial whaling will continue indefinitely.[94] To this end, the IWC has markedly changed from a conservationist into a preservationist body,[95] and the objective of the ICRW regarding sustainable whaling and the organisation of the whaling industry has accordingly changed into preservation of whales for future present and future generations.[96]

However, the fact remains that the IWC has continuously increased the allocation of ASW quotas. Awareness concerning the welfare of whales and the growing support for the movement of rights of whales which is a part of the rights approach to animals may therefore be particularly important in this regard.[97] Promoting whale welfare has been a continuously expanding activity of the IWC. Already before the adoption of the moratorium on commercial whaling, the IWC had started to occupy itself with the 'humane killing' of whales. Since 1981, the use of the cold grenade harpoon for killing for commercial purposes has been prohibited.[98] In 1999, the IWC established a working group on Whale Killing Methods and Welfare Issues. The mandate of this working group is today to review information and provide advice to the Commission on issues relating to whale killing methods and all aspects associated with ensuring good welfare of cetaceans.[99]

Despite the availability of modern weaponry and training of indigenous hunters by whalers from the US and Norway, outdated equipment and methods are still used.[100] Therefore, an even stronger protection of whales would come through an acknowledgement of whales' rights as opposed to whale welfare.[101] Whales' rights comparable to human rights had already been asserted by some participants at a conference on the non-consumptive use of cetaceans in Boston in 1993. An interdisciplinary group of scholars, in 2010, adopted a 'Declaration of Rights of Cetaceans: Whales and Dolphin'. The preamble notes that 'the progressive development of international law manifests an entitlement to life by cetaceans'. The idea of animal rights is gaining traction. It may well be that the debate about a whale right to life will be reinvigorated by the new domestic case law on habeas corpus.[102] The ultimate goal would be granting the right to life to whales, but it seems unlikely at present. As it stands now, the functioning of the IWC has become less conformational and it will influence favourably and effectively the preservation of whales for present and future generations. The right to self-determination, when extended to indigenous peoples, means that indigenous peoples should be free to decide on the development of their culture. This should not, however, under any circumstances mean that indigenous peoples are free to engage in environmentally dangerous practices. If they are to be subjects of international law, they must necessarily be bound by the same environmental principles – such as the sustainability requirement and precautionary principle – as States.[103]

Accordingly, indigenous whaling should in principle conform, if not to animal rights approaches, at least to animal welfare approaches, that is to say, to ensure that pain and suffering is minimised. To this end, there have been a number of cases in which the UNHRC has clarified the rights of indigenous peoples regarding Article 27 ICCPR,[104] addressing, in particular, the issue of the use of modern technology within traditional indigenous hunting activities. Thus, in the 1994 *Länsman* case, the UNHRC stated that modern practices adopted by indigenous peoples did not prevent them from invoking Article 27 ICCPR.[105] In other words, Article 27 ICCPR allows for adaptation of those means to the modern way of life and its ensuing technology.[106] However, the use of traditional methods of killing whales by indigenous peoples is a very sensitive issue and of great concern. The contemporary approach of the IWC to the use of technology follows the stand adopted by the UNHRC.[107] However, this is not a uniformly accepted position, as the UNHRC acknowledgement of the evolving lifestyle of indigenous peoples and also their permissible use of modern technology is seen by certain States as a contentious issue within the IWC; whereas improved technology makes whaling safer, more efficient and more humane, it may be seen by some to compromise aboriginal authenticity.[108]

Therefore, as the practice of the IWC indicates, what it understands by culture and the method of ASW is not entirely clear. That said, it may be difficult to object to ASW provided that: (a) such whaling fulfils a legitimate and continuing cultural, nutritional, and subsistence need; (b) such killing is limited to only the number of whales needed; (c) the targeted whale populations can sustain such kills; (d) each whale is fully utilised by those responsible for the animal's death and not traded commercially; (e) the whaling is conducted using the least cruel techniques available; and (f) continuing efforts are made to increase the efficiency of the hunt and reduce the amount of time it takes individual whales to die. The regulation of ASW – not least as concerns maximum quota, humane killing methods using modern technologies, and sustainability – is therefore likely to be a key element in the future of the IWC agenda.

Notes

1 At the time, the Ministry of Foreign Affairs of Japan stated that: "From July 2019, after the withdrawal comes into effect on June 30, Japan will conduct commercial whaling within Japan's territorial sea and its exclusive economic zone, and will cease the take of whales in the Antarctic Ocean/the Southern Hemisphere. The whaling will be conducted in accordance with international law and within the catch limits calculated in accordance with the method adopted by the iwc to avoid negative impact on cetacean resources". See Ministry of Foreign Affairs of Japan, Statement by Chief Cabinet Secretary, 26 December 2018, <https://www.mofa.go.jp/ecm/fsh/page4e_000969.html> accessed 4 August 2022.
2 International Convention for the Regulation of Whaling 1946. The Japanese government justified the resumption of commercial whaling by asserting that the IWC has "refused to agree to take any tangible steps towards reaching a common position that would ensure the sustainable management of whale resources", see ibid. For an overview of the consequences of Japanese withdrawal, see Nikolas Sellheim, *International Marine Mammal Law* (Springer 2020).
3 Norway objected, Iceland left the Convention but re-joined with a reservation; See Couzen's chapter in this volume.
4 Paragraph one reads as follows: '1. Notwithstanding anything contained in this Convention any Contracting Government may grant to any of its nationals a special permit authorizing that national to kill, take and treat whales for purposes of scientific research subject to such restrictions as to number and subject to such other conditions as the Contracting Government thinks fit, and the killing, taking, and treating of whales in accordance with the provisions of this Article shall be exempt from the operation of this Convention. Each Contracting Government shall report at once to the Commission all such authorizations which it has granted. Each Contracting Government may at any time revoke any such special permit which it has granted'.
5 *Whaling in the Antarctic (Australia v Japan: New Zealand intervening)* (Judgment) ICJ Rep 2014 p 226.
6 See map in Sellheim's and Morishita's chapter in this volume.
7 See for instance Doug Bock Clark, *The last whalers: Three years in the far pacific with a courageous tribe and a vanishing way of life* (Little, Brown and Company 2019).
8 The Indigenous methods of whaling has given rise to questions regarding cruelty, an issue which can also be analysed within the paradigm of animal rights. '... ASW tend to be less efficient than those used in commercial whaling operations, with the result that (a) times to death are longer, (b) instantaneous death rates are lower, and (c) "struck and lost" rates are higher. Although the IWC has adopted several resolutions seeking improvements in the humaneness of ASW operations, IWC resolutions are not binding on parties. 'Aboriginal Subsistence Whaling', Animal Welfare Institute', https://awionline.org/content/subsistence-whaling accessed 3 June 2023.
9 International Covenant for Civil and Political Rights (ICCPR), 1966.
10 UN Declaration on the Rights of Indigenous Peoples, 2007.
11 Anne Peters, *Animals in International Law*, Académie de Droit International de la Haye, Offprint from Collected Courses of The Hague Academy of International Law – Recueil des cours, Volume: 410 (Brill Nijhoff 2020) 201–203.
12 G P Donovan, 'The International Whaling Commission and aboriginal/subsistence whaling: April 1979 to July 1981', in G P Donovan (ed.), *Aboriginal/subsistence*

whaling (with Special Reference to the Alaska and Greenland Fisheries (Reports of the International Whaling Commission Special Issue 4, Cambridge 1982) 79, 83.

13 Athanasios Yupsanis, 'Article 27 of the ICCPR Revisited: The right to culture as a normative source for minority/indigenous participatory claims in the case law of the human rights committee', (2013), *26 Hague Yearbook of International Law* 359, 410.

14 See further Junichi Takahashi, Arne Kalland, Brian Moeran and Theodore C. Bestor, 'Japanese whaling culture: Continuities and diversities', (1989), *Maritime Anthropological Studies* 2(2), 128.

15 See General Comment 21, paras 36 and 37.

16 Preamble of the 2007 United Nations Declaration on the Rights of Indigenous Peoples, adopted 13 September 2007.

17 See e.g. '[The] shift away from positivist, state dominated dialogue toward a more inclusive framework that is much more responsive to the ideals enshrined in the Charter of the United Nations ... has created [for indigenous peoples] a space for them to move an agenda of promoting and encouraging respect for their human rights within this formal international organization, including the collective rights to their culture, their land, and self-government as an essential part of their individual self-realization'. International Law Association, The Hague Conference 2010, Rights of Indigenous Peoples, Professor Siegfried Wiessner, Chair; Dr Federico Lenzerini (Italy), Rapporteur, 3.

18 Article 3: 'Indigenous peoples have the right to self-determination. By virtue of that right they freely determine their political status and freely pursue their economic, social and cultural development.'

19 Article 4: 'Indigenous peoples, in exercising their right to self-determination, have the right to autonomy or self-government in matters relating to their internal and local affairs, as well as ways and means for financing their autonomous functions.'

20 Wiessner (n) 143.

21 *Kitok v. Sweden,* Comm. No. 197/1985, Supplement No. 40 (A/43/40), 221–230, 27 July 1986; CCPR/C/33/D/197/1985.

22 *Ominayak, Chief of Lubicon Lake Band v. Canada*, Comm. No. 267/1984. Report of the HRC, UN GOAR, 45th Sess., Supp. No. 2 at 1.

23 James Anaya, 'International human rights and indigenous peoples: The move toward the multicultural state', (2004), 21 *AJICL* 9, 29.

24 *Lansmänn and Others v. Finland*, Comm. No. 511/1992, HRC, UN Doc. CCPR/C/52/D/ 511/1992 (8 November 1994).

25 Malgosia Fitzmaurice, *Whaling in International Law* (CUP 2015).

26 Will Kymlicka, *Multicultural Odysseys: Navigating the new international politics of diversity* (OUP 2007) 149.

27 ibid.

28 Convention for the Regulation of Whaling 1931.

29 It may be noted that in the Arctic context both sails and motors arrived around the same time and are exogenous technologies embraced, along with firearms, by indigenous peoples. This approach when considered in relation to Article 3 of the GWC may indicate why motorised whale craft are in some cases perceived as commercial craft (ie, in Japan), but not others (Inupiat whale boats).

30 Protocol amending the International Agreement of 8 June 1937 for the Regulation of Whaling (entered into force 24 June 1938) 196 LNTS 131. For the Parties to the 1931 Convention, Article 3 was in force.

31 See Hisashi Hamaguchi, 'Aboriginal subsistence whaling revisited', in Nobuhiro Kishigami, Hisashi Hamaguchi and James M. Savelle (eds.), *Anthropological studies of whaling* (Senri Ethnological Studies 2013) 81—82.

32 ibid.

33 Report of the Cultural Anthropology Panel, reprinted in: International Whaling Commission, 'Aboriginal/Subsistence Whaling (with special reference to the Alaska and

Greenland fisheries)', (1982), Reports of the International Whaling Commission – Special Issue 4, 35–50, <www.iwcoffice.org/cache/downloads/ebvrl7xp4e80w40804c-ssc4s8/RIWC-SI4-p p34-73.pdf> accessed 4 August 2022. It may be observed that this definition which at present is exclusively applicable to Indigenous peoples, could also include Japan even considering that coastal whalers provide whale meat to urban residents in cities as large as Tokyo who share the tradition.

34 Randall Reeves, 'The origins and character of "Aboriginal Subsistence" whaling: A global review', (2002), *Mammal Review* 32, 71, 77.

35 Hamaguchi (n) 86.

36 ibid.

37 ibid.

38 Reeves (n) 77.

39 Alexander Gillespie, 'Aboriginal subsistence whaling: A critique of the interrelationship between international law and the IWC', (2001), 12 *CJIELP* 77.

40 Heinämäki (n) 45.

41 Ibid., 44.

42 See above.

43 Heinämäki (n) 45

44 IISD, 'IWC report details recommendations on aboriginal subsistence whaling,' 20 October 2015, <http://sdg.iisd.org/news/iwc-report-details-recommendations-on-aboriginal-subsistence-whaling> aaccessed 4 August 2022.

45 IWC, 'Aboriginal subsistence whaling', <https://iwc.int/aboriginal> accessed 4 August 2022. It may be noted that the modern definition of ASW is consistent with the modern whaling practices of Norway, Iceland and Japan. It may open a possibility of merging Japanese coastal whaling with ASW.

46 ibid.

47 ibid.

48 ibid.

49 IWC, Resolution 1981–4: Resolution to the Government of the United States on the Behring Sea Bowhead Whale.

50 Richard Caddell, 'Marine mammals at the poles', in Karen N. Scott and David L. VanderZwaag (eds.), *Research handbook on polar law* (Edward Elgar 2020) 241; IWC, Resolution 1998–9: Resolution on Directed Takes of White Whales; Resolution 1994–4: Resolution on a review of aboriginal subsistence whaling.

51 IWC, Resolution 1993–2: Resolution on Pilot Whales; IWC, Resolution 1995–1: Resolution on Killing Methods in the Pilot Whale Drive Hunt.

52 International Whaling Commission, https://iwc.int/aboriginal accessed 4 August 2022.

53 https://lauda.ulapland.fi/bitstream/handle/10024/63587/Sellheim.Nikolas.pdf?isAllowed=y&sequence=1 accessed 3 June 2023.

54 Malgosia Fitzmaurice, 'Indigenous peoples in marine areas: Whaling and sealing', in Stephen Allen, Nigel Bankes and Øyvind Ravna (eds.), *The rights of indigenous peoples in marine areas* (Hart Publishing 2019) 85.

55 See below.

56 Fitzmaurice, 'Indigenous Peoples in Marine Areas: Whaling and Sealing', (n) 85–88.

57 Hilde Woker, 'The rights of indigenous peoples to harvest marine mammals in the arctic', (2015). <https://munin.uit.no/bitstream/handle/10037/8468/thesis.pdf?sequence=2> accessed 21 April 2023, 36.

58 IWC, Report of the Thirty-Fourth Meeting, Chair ' s Report (1983), 38 Appendix 3, cited in Woker (n) 37.

59 Woker (n) 37.

60 Ibid., 38.

61 See discussion in Fitzmaurice, 'Indigenous Peoples in Marine Areas: Whaling and Sealing' (n).

62 In Canada Aboriginal whaling takes place in the Canadian Arctic. Canadians kill about 600 narwhals per year. They kill 100 belugas per year in the Beaufort Sea, 300 in northern Quebec (Nunavik), and an unknown number in Nunavut. The total annual kill in Beaufort and Quebec areas varies between 300 and 400 belugas per year, see Randall Reeves and David S Lee, 'Bowhead whales and whaling in the central and eastern Canadian arctic, 1970–2021', (2022), *Journal of Cetacean Research and Management* 23(1), 1.

63 Heinämäki (n) 46–52.

64 See   https://www.washingtonpost.com/nation/2021/09/29/makah-whaling-judge-recommendation/ accessed 4 August 2022.

65 See further https://www.nationalgeographic.com/environment/article/us-tribe-wants-to-resume-whale-hunts-will-conservationists-support-them accessed 4 August 2022.

66 See Nobuhiro Kishigami, 'Aboriginal subsistence whaling in Barrow, Alaska' in Kishigami, Hamaguchi and Savelle (n) 101, 116.

67 The Greenland Home Rule Executive Order No. 2 of 12 February 2004.

68 Frank Sejersen, 'Hunting and management of beluga whales (Delphinapterus leucas) in Greenland: Changing strategies to cope with new national and local interests', (2001), *Arctic* 54, 431, 434.

69 Chris Wold and Michael S. Kearney, 'The legal effects of Greenland's unilateral aboriginal subsistence whale hunt', (2015), 3 *AUILR* 561, 603.

70 ibid.

71 Hardy Jones, 'Greenland begins humpback whale hunt' (New York City, Huffington Post, 24 August 2010) <www.huffingtonpost.com/ hardy-jones/greenland-begins-humpback_b_693054.html> accessed 21 April 2023.

72 John D, 'The whale as Kami', available at https://www.greenshinto.com/2013/08/14/the-whale-as-kami/ accessed 3 June 2023.

73 Junichi Takahashi, Arne Kalland, Brian Moeran and Theodore C. Bestor, 'Japanese whaling culture: Continuities and diversities', (1989), *Maritime Anthropological Studies* 2(2), 128.

74 See IWC, Resolution 1995–8: Resolution on Whaling under Special Permit in Sanctuaries; IWC, Resolution 1998–3: Resolution on the Southern Ocean Sanctuary. The IWC adopted a Resolution 2014–5 on scientific whaling, which aimed at the implementation of the ICJ's proportionality test and instructed the Scientific Committee to review all scientific permits (new and previously issued) and the necessity of the lethal methods in order to achieve scientific objectives (Resolution 2014–5, 18 September 2014 on Whaling under Special Permit). In 2016, the IWC adopted the Resolution 2016–2 which stated that Japan had started NEWREP-A before the IWC's assessments (Resolution 2016–2, 28 October 2016 on Improving the Review under Special Permit). In 2017, the IWC's report stated that the NEWREP-NP does not justify taking of lethal samplings (Report of the Expert Panel for New Scientific Whaling Research Programme in the Western North Pacific (NEWREP-NP, 3 March 2017, SC/^&/A/REP/01, p.44). EU, represented by the Netherlands, encouraged Japan to follow the IWC's procedures which were set out in 2014 and 2016 Resolutions ('Whaling under Special Permit: Letter from the Netherlands of behalf of the EU Member States party to the ICRW', 18 January 2017 (SB/JAC/32251)). During the session at Florianópolis, the IWC assessed the NEWREP-A as not conforming to the criteria of special permits under Article VIII (Annex O of the IWC, Report of the 67th Meeting). In the Florianopolis Declaration, the lethal scientific killing was condemned as unnecessary (Florianopolis Declaration). In light of this, Japanese scientific whaling for the season 2019/2020 was subsumed by and merged with resumed commercial whaling (statement of the chief cabinet secretary, 26 December, 2018, <https://www.mofa.go.jp/ecm/fsh/page4e_000969.html> accessed 21 April 2023).

75 The 68th meeting of the IWC was due to take place in 2020 but was postponed to 2022 due to the Covid-19 outbreak.

76 See above.

77 Peters (n) 186–194.

78 ibid. Stone argues that the system of allocation of whale quotas for Aboriginal whaling leads to unexpected results, such as the Chukchi people feeding farmed foxes with allocated gray whale meat. He also presents a very interesting analysis of the Makah people resumption of whaling, asking if this is really good to step backwards (ibid., 283).

79 Hamaguchi (n) 152.

80 'More than 800 whales were condemned today just in the Greenland vote', Wendy Higgins of the Humane Society International (HSI) told Agence France Press (AFP) on the first day of the controversy-laden gathering in Slovenia. 'We are concerned that the new IWC quota will give Greenland more whale meat than its native people need for nutritional subsistence and that the surplus will continue to be sold commercially, including to tourists', said the Animal Welfare Institute (AWI). See Celine Serrat, 'Whaling: Greenland Hunt Gets Okay, Iceland Blasted' (Yahoo News, 15 September 2014), <https://sg.news.yahoo.com/whale-huddle-braces-clash-over-japanese-hunting-004713650.html> accessed 21 April 2023).

81 According to the AWI: 'When the non-governmental organizations were allowed to speak, the chair called on Whale and Dolphin Conservation, which spoke on behalf of itself and AWI and challenged the tonnage of whale meat that Greenland seeks because it is significantly more than all the marine protein currently being consumed by Greenlandic people, from whales, seals and small cetaceans put together' (AWI, '2014: IWC65 in Portoroz, Slovenia', <https://awionline.org/content/2014-iwc-65-meeting-slovenia> accessed 21 April 2023).

82 IISD, 'Summary report, 10–14 September 2018' <https://enb.iisd.org/events/67th-meeting-international-whaling-commission-iwc67/summary-report-10-14-september-2018> accessed 4 August 2022.

83 Caddell (n) 242; IWC, Annual report of the International Whaling Commission 2012 (IWC, Cambridge, 2013), 22–24.

84 Martin Hennig and Richard Caddell, 'On thin ice? Arctic Indigenous communities, the European Union and the sustainable use of marine mammals', in Liu, Kirk and Henriksen (n) 296, 328–334.

85 ibid.

86 See AWI, 'AWI opposes judge's recommendation to allow Makah tribe to hunt gray whales' <https://awionline.org/press-releases/awi-opposes-judges-recommendation-allow-makah-tribe-hunt-gray-whales> accessed 21 April 2023.

87 AWI, 'Aboriginal subsistence whaling' <https://awionline.org/content/subsistence-whaling> accessed 21 April 2023.

88 In particular, a cold harpoon (banned by the IWC for commercial whaling and an eight-foot lance is repeatedly stabbed at the animal in an attempt to puncture the whale's heart or lungs. In some cases, the whale is finally killed by a "bomb lance" – an exploding projectile discharged from a shoulder gun. 'Aboriginal Subsistence Whaling', Animal Welfare Institute, <https://awionline.org/content/subsistence-whaling#:~:text=Most%20minke%20whales%20and%20all,primary%20and%20secondary%20killing%20methods> accessed 3 June 2023. See also Anne Peters (ed.), *Studies in global animal law* (Springer 2020).

89 Milton R Freeman, 'Is Money the root of the problem? Cultural conflicts in the IWC', in Robert L Freidheim (ed.), *Toward a sustainable whaling regime* (UWP, 2001) 129.

90 North Atlantic Right Whale – Critically Endangered; Vaquita – Critically Endangered ; Atlantic Humpback Dolphin – Critically Endangered; North Pacific Right Whale – Endangered; Sei Whale – Endangered; Blue Whale – Endangered; Western Gray Whale – Endangered, see https://www.treehugger.com/most-endangered-whales-on-earth-4863926 accessed 3 June 2023.

91 Evidenced by the *Bering Sea Arbitration, United Kingdom v United States*, (Award, (2007) XXVIII *RIAA* 263, (1898), 1 *Moore's International Arbitrations* 755, 15 August 1893).

92 The 68th meeting of the IWC was due to take place in 2020 but was postponed to 2022 due to the Covid-19 outbreak.

93 See Preamble recitals.

94 IWC, Chair's Report of the 67th Meeting, Rep2018 (Chair 67), September 2018; and IWC, 'The Florianópolis declaration on the role of the International Whaling Commission in the conservation and management of whales in the 21st century', Res 2018–5, September 2018).

95 See *Jefferies' and Stock's chapter in this* volume.

96 Fitzmaurice, Whaling in International Law (n) 86–87).

97 Peters (n) 186–194. In line with this approach is the 2010 Declaration of Rights of Cetaceans: Whales and Dolphins (2010 Cetaceans Rights: Fostering Moral and Legal Change'.

98 Article 6 of the Schedule.

99 Peters (n) 228–229.

100 https://awionline.org/content/subsistence-whaling.

101 See in this regard, Russel Lawrence Barsh, 'Indigenous peoples', in Daniel Bodansky, Jutta Brunnée and Ellen Hey (eds.), *The Oxford handbook of International Environmental Law* (OUP 2007) 850; Heinämäki, 'Protecting the rights of indigenous peoples – Promoting the sustainability of the global environment?', (2009), 11 International *Community Law Review* 3, 66; Nancy C. Doubleday, 'Aboriginal subsistence whaling: The right of inuit to hunt whales and implications for International Environmental Law', (1988–1989) 17 *Denver Journal of International Law & Policy* 17 373.

102 Peters (n) 230–231.

103 Benjamin J. Richardson, 'Indigenous peoples, international law and sustainability', (2001), 10 *RECIEL* 1, 3.

104 See e.g. HRC, *Ilmari Länsman et al. v. Finland,* Views adopted on 26 October 1994, Comm. No. 511/1992, UN Doc. CCPR/C/57/1, 75–85; *Jouni E. Länsman et al. v. Finland,* Views adopted on 30 October 1996, Comm. No. 671/1995, UN Doc. CCPR/C/58/D/671/1995.

105 *Jouni E. Länsman et al. v. Finland*, para. 9.3.

106 *Apirana Mahuika et al. v. New Zealand*, Views adopted on 27 October 2000, Comm. No. 447/1993, UN Doc. CCPR/C/70/D/547/1993, para. 9.4.

107 Report of the Panel Meeting of Experts on Aboriginal/Subsistence Whaling, reprinted in: International Whaling Commission, 'Aboriginal/subsistence whaling (with special reference to the Alaska and Greenland fisheries)', Reports of the International Whaling Commission – Special Issue 4 (1982), 7–9, <http://iwcoffice.org/cache/down loads/7h78clbatfggcw8gg00w8wo4o/RIWC-SI4-ppl-33.pdf> accessed 4 August 2022. See also Randall R. Reeves, 'The origins and character of "Aboriginal Subsistence" whaling: A global review', (2002), *Mammal Review* 32, 71, 98.

108 International Whaling Commission, 'Aboriginal subsistence whaling' <http://iwc.int/aboriginal> accessed 4 August 2022.

8 Canada's Withdrawal from the International Whaling Commission in 1982 and Its Restoration of Inuit Bowhead Whaling. Lessons for Japan's Restoration of Coastal Whaling?

Barry Scott Zellen

Past as Prologue

Japan's historic withdrawal from the IWC announced the day after Christmas in 2018 and becoming effective on 30 June 2019 follows 37 years after another historic exit from the IWC for parallel reasons, even if this might seem paradoxically so: that of Canada, announced – like Japan's IWC exit – to global headlines in the summer of 1981 (before becoming effective 30 June 1982). In the years since, Japan has refocused its whaling efforts on coastal commercial whaling within its own Exclusive Economic Zone (EEZ), having brought to an end its long-running global scientific whaling program from the Southern Ocean to the High North Pacific. Japan's transition from global scientific whaling on the high seas to regional coastal whaling in one's own territorial seas mirrors the evolution of whaling elsewhere, notably the transition from no large-cetacean subsistence whaling by Canadian Inuit to what has now become a fully revitalized, persistent and continuous annual subsistence whaling tradition in Arctic Canada, and also with parallels to contemporary commercial coastal whaling in Iceland conducted side-by-side with and from the same ports as whale-watching cruises.[1]

In each of these cases, the evolution and modernization of whaling in these two very different maritime nations has followed its own distinct path – one channelled through the revitalization of age-old indigenous cultural practices of subsistence harvesting, the other through a locally revitalized commercial whaling industry. But each transporting the culture and economy of whaling and its regulatory framework from the contested realm of international politics[2] to a more welcoming and supportive domestic politics context where external actors have little or no influence. Japan is thus now commencing its fifth season of commercial coastal whaling since exiting the IWC in 2019, following a path which was in many ways blazed a generation ago in northern Canada, albeit for subsistence and not commercial whaling purposes – but each embracing their whaling heritage and rightsizing its practice primarily for domestic consumption and regulation. The purpose of this chapter is to compare Japan's current transition in whaling and Canada's, both precipitated by their departures from the IWC, and whilst heading in different

DOI: 10.4324/9781003250814-10

directions (the latter for purposes of indigenous subsistence, the former commercial), their journeys nonetheless possess much in common.

A look back at Canada's 1982 IWC exit and its consequences, particularly the renaissance in subsistence coastal whaling by Canadian Inuit that gradually and with progressively increasing vigor followed – steadily gaining momentum while increasingly expanding across the entirety of the Canadian Arctic in the four decades since – will shed important light in what we may expect for Japanese coastal whaling in the years ahead. Indeed, with its fifth summer of coastal whaling under way at the time of this chapter's completion since quitting the IWC, Japan's centuries-long, and indelibly proud cultural whaling tradition navigates its way from the rough and tumble Hobbesian international arena,[3] to a more Rousseauian domestic context where whaling is effectively shielded from external interference by the Law of the Sea so long as it is practiced inside one's EEZ.

Insulated from the perils of international anarchy, Japanese whalers can refocus on overcoming the declining mainstream Japanese interest in consuming whale meat. This has continued even with the IWC exit and shift to coastal whaling, amidst persistent criticism from international environmental activists,[4] as Japanese coastal whalers work to preserve Japan's centuries-long cultural tradition of whaling on a more local scale; conducted primarily by smaller and less expensive to operate catcher vessels kitted out with elevated sighting platforms and a harpoon gun, as the industry's ageing flagship, the now 35-year old factory whaler *Nisshin Maru* – no longer wears signs and banners proclaiming its status as a research vessel and instead plies Japanese coastal waters with its more nimble brethren, awaiting its successor to be constructed. The secular decline in whale consumption by the Japanese populace may prove the biggest challenge to this effort at cultural revitalization, as whale meat, long a staple and in the immediate aftermath of World War II a vital path towards a return of Japan's nutritional self-sufficiency (one strongly supported by the U.S. military government administering the post-conflict nation) continues to fall out of favor, but creative efforts are under way to spark renewed popular support for Japan's whaling pastime.

In Canada, bowhead whaling is practiced for subsistence and symbolic purposes with passion and enthusiasm by Inuit whalers and their communities, with each successful hunt precipitating a community feast and an intercommunal sharing of whale meat that celebrates the proud restoration of a cultural tradition that only recently faced the same oblivion that threatens Japan's struggling whaling industry. The Canadian example of a dramatic IWC exit paving the way to a successful restoration of coastal whaling across Canada's Arctic territories offers much insight and understanding in the successful revitalization of a whaling culture amidst a national rescue and resuscitation effort linked to the disentanglement of whaling as a national cultural prerogative from the messy and chaotic dynamics of international politics. In addition, it now occurs independently from the paradoxical internal dynamics of the IWC which has for so long been held hostage by anti-whaling sentiment espoused oftentimes by former whaling nations who are themselves responsible for the decimation of whale stocks during the colonial era.

1982: Ottawa Bids Farewell to the IWC

On 28 June 1981, the *New York Times* reported that Canada had announced it would withdraw from the IWC because of a lack of "any direct interest in the whaling industry or in the related activities of the commission," having one decade prior "outlawed commercial whaling in 1972."[5] According to Andrew P. Hutton, Ottawa's decision was not universally popular; Prime Minister Pierre Trudeau explained that Ottawa "was under heavy pressure from environment groups who blamed it for decisions taken by whaling nations which Canada did not agree with" and noted that Canada would "still be able to express whatever arguments we can within the commission" while no longer being "part of the decisions by those countries which continue to be whaling nations."[6] But opposition New Democratic Party (NDP) spokesman Ian Deans countered

> that if every other nation in the IWC took the same position, only the countries which have whaling fleets would be involved in the conservation decisions in order to protect the dwindling whale pods. Surely it would make more sense for the government to continue to be at the bargaining table, to continue to have a vote and an opportunity to influence the outcome rather than to abdicate responsibility, and ultimately destroy the species, Deans said.[7]

Echoing the NDP, the *New York Times* reported that the "Whale Project, a lobbying coalition representing 30 environmental groups, protested the withdrawal, which it called a regrettable abdication of responsibility for the survival of the world's whale population."[8]

But as would be increasingly recognized in the coming decades, not just by federal officials but also by indigenous subsistence whalers, their families and their communities that co-managed the reintroduction of large cetacean harvesting for subsistence purposes, subsistence bowhead hunting in Canada presented a wildly successful test-case of collaborative governance under new land claim agreements across the Inuit homeland. These were sequentially achieved in the wake of the repatriation from Britain of the Canadian Constitution in 1982 just weeks before Ottawa's IWC exit came into effect, from the 1984 Inuvialuit Final Agreement (IFA) through the 1993 Nunavut Land Claims Agreement and subsequent formation of the Nunavut Territory in 1999, the 2005 Labrador Inuit (Nunatsiavut) Land Claims Agreement, and the 2007 Nunavik Inuit Land Claim Agreement – each reinvigorating through the formal protections provided of indigenous subsistence, inclusive of and symbolically embodied by bowhead whaling, upon their respective settlement. While the earlier Inuit bowhead whale hunts generated vociferous protest by anti-whaling activists, in time even organizations like Greenpeace came to see indigenous whaling as acceptably sustainable and to respect the indigenous rights being protected thereby. Though it would be a slow, decades-long evolution that witnessed Greenpeace evolving from one of the most widely reviled organizations among Inuit hunter into a now-apologetic supporter of Inuit whaling, already there is much reason to hope for a more rapid reconciliation between Japan and its whaling industry on the one hand, and the international environmentalists and anti-whaling organizations like Greenpeace and Sea Shepherd that have long vocally opposed Japanese whaling.

Curiously, and importantly, the 30 June 1982 IWC exit by Ottawa followed quite rapidly upon the historic 1982 repatriation of the Canadian Constitution on 17 April of that year, though the decision to withdraw from the IWC was made the prior year ahead of the repatriation.[9] The 1982 Constitution Act not only enabled amending the Constitution without Britain's consent, thus bringing a close to Britain's long colonial project in North America, but also enshrined the Canadian Charter of Rights and Freedoms, which included Section 35 affirming the rights of indigenous peoples (including, specifically, those "that now exist by way of land claims agreements or may be so acquired"[10]) in the Canadian Constitution, with a profound impact on the historic reconciliation of Canada's indigenous peoples with Ottawa, and the historic re-empowerment of indigenous peoples through land claims settlement and self-government agreements in the four decades that have followed.[11] As Robert Sheppard described on *CBC*:

> [W]hen it comes to the IWC we have turned aside repeated entreaties even by close allies like the U.S. to join and help them try to keep a lid on commercial whaling. The history here is that Canada quit the IWC in 1982 as it was adopting its moratorium on commercial whaling (the full ban went into effect four years later) and also attempting to regulate the number of whales that could be hunted by aboriginal groups. At the time Canada's Inuit in the eastern Arctic wanted to resume hunting bowheads, but the IWC saw this particular group of whales (there were only about 500 of them in that part of the world) as endangered and would not go along. The politics here was that the Canadian Constitution had just been patriated through an arduous 18-month political battle and the northern Inuit had been among the Trudeau government's most loyal native allies in that fight. As part of a promise to them, the Trudeau government withdrew from the IWC, banned commercial whaling in its territorial waters, and set its own quotas for aboriginal whaling. Subsequent Canadian governments have stayed true to that promise ever since. Is it a good bargain? Canadian Inuit have not been what you would call conspicuous whalers. Between 1982 and 1996 they only took one bowhead from the eastern Arctic. By contrast, Alaskan Inuit in the western Arctic, where there are many more of these big beasts, were averaging 50 a year. In the summer of 1996, the Inuit hunted two bowheads. That earned them the condemnation of the IWC and only reinforced Canada's resolve to stay out of that body altogether.[12]

1991-Present: A Renaissance in Inuit Coastal Whaling

While Ottawa's 1982 IWC exit was not initially framed as a component of this constitutional repatriation, the historical convergence of these two sequential events resulted in indigenous coastal whaling communities being among the most visible beneficiaries of Ottawa's decision to withdraw from the IWC. This fostered a renaissance on indigenous coastal whaling that began in 1991 with the Inuvialuit bowhead hunt in September of that year, and which continued with over three dozen subsequent Inuit bowhead hunts in the new territory of Nunavut. Its 1999 formation was agreed to in the 1993 Nunavut land claim accord, as well as Nunavik in northern Quebec, to Nunavut's immediate south, which in 2007 extended its 1975 land claim agreement to include marine resources off northern Quebec's shores.

The history of Inuit land claims in Canada and the protection that land claims agreements afforded Inuit whaling are recounted by Randall Reeves and David Lee in the IWC's *Journal of Cetacean Resource Management*, with several milestones in their journey noted, including a pivotal 1979 resolution by the Inuit Circumpolar Council[13] that proclaimed "whaling is a necessary part of Inuit cultural identity and social organisation, and is in no way similar to commercial whaling." That called upon the IWC to "defend Inuit rights to hunt the whale," upon Inuit to make "wise and full use of subsistence resources," and upon Arctic nations to "specifically provide for the determination of safe technology; Arctic population policy; and locally controlled wildlife management."[14] Reeves and Lee also noted the 1975 James Bay and Northern Québec Agreement (JBNQA) with its "framework within which the rights of Inuit to hunt whales (and other wildlife) 'sustainably' was guaranteed" as well as the 1984 IFA "stipulating, among other things, that the Inuvialuit have the right to hunt all marine mammals for subsistence" with catches limited by quotas "set jointly by the Inuvialuit and the Government according to the principles of conservation."[15]

Prior to the landmark passage of the IFA, "Inuvialuit hunters had attempted to organise a bowhead hunt, but without success"; afterwards, the very "first successful legally sanctioned hunt for bowheads in the western Canadian Arctic for many decades took place in 1991, with one whale taken and butchered at Shingle Point, NWT"; while Ottawa "continued to issue licences to the Inuvialuit to take one bowhead per year [...] none were landed until July 1996, again at Shingle Point."[16] The pioneering 1991 bowhead hunt, while one of only two conducted by the Inuvialuit in an affirmation of their 1984 land claim, provided an important and historic catalyst for the resumption of bowhead hunting across the Canadian Arctic. And just as those first two hunts celebrated the restoration of Inuit hunting rights inclusive of large cetaceans like bowheads, the next three dozen hunts would affirm and celebrate subsequent Inuit land claim accords, next in Nunavut (signed in 1993), and later in Nunavik, which extended in 2007 the JBNQA to include "marine waters bordering far northern Labrador and northern Quebec, including large offshore islands in Ungava Bay, Hudson Strait, and northeastern Hudson Bay"[17] that had been hitherto excluded from Inuit resource co-management in 1975.

According to Reeves and Lee, the 1993 Nunavut Land Claim Agreement constitutionally protected Inuit "wildlife harvesting rights and rights to participate in decision-making concerning wildlife harvesting," in part so as to "encourage self-reliance and [...] cultural and social well-being," and thus

legally sanctioned hunting of bowheads in Nunavut (i.e., approved by the federal Minister of Fisheries and Oceans and authorised by the Nunavut

Wildlife Management Board, NWMB, the primary instrument of wildlife management in Nunavut) began with the successful landing of a large whale by the Inuit of Naujaat in 1996.[18]

Similarly, the 2007 Nunavik Inuit Land Claims Agreement, signed in 2007, "established the Nunavik Marine Region Wildlife Board (NMRWB), which is similar in composition and function to the NWMB. The first bowhead hunted under the terms of the agreement was taken at Kangiqsujuaq on 9 August 2008."[19]

Reeves and Lee also chronicle the geographic expansion of bowhead hunting and concomitant increase in bowhead quotas by Canadian Inuit:

Before 2008, the total allowable harvest' (TAH) in Nunavut was one whale every 2–3 years as established through the Nunavut Agreement. In 2008 it was increased to 2 per year and in 2009 to 3 per year for the next 3 years. [...] In 2014 the TAH for Nunavut was increased to 4 and in 2015 to 5. The NMRWB has established a total allowable take (TAT) of 2 bowheads per year for the Nunavik Marine Region.[20]

These quotas have held steady since then, and to date, there has been on average one bowhead hunt per year since Ottawa's exit from the IWC just over 40 years ago, with each hunt serving as a powerful symbol of the revitalization of Inuit culture achieved since the Constitution Act was passed in 1982. By Reeves' and Lee's count, 37 bowheads (16 males, and 21 females)[21] were harvested by Canadian Inuit between 1994 and 2020, and with the 1991 hunt at Shingle Point by the Inuvialuit and the 2021 hunt in Repulse Bay by the Inuit of Nunavut added, this totals 39. As Reeves and Lee observe,

Since the mid 1990s, the hunting of bowhead whales in the central and eastern Canadian Arctic has become an increasingly regular feature of the annual round in several Inuit communities. Sixteen communities have landed at least one bowhead over the past 27 years, with 9 of them landing 2 or more.[22]

Two or three unauthorized hunts were also chronicled over this same time period – one unconfirmed kill in Nunavik in 1979; and a confirmed kill at both Arviat in 1985 and Igloolik in 1994, both in the Nunavut Territory – bringing the total number of bowheads hunted by Canadian Inuit since bowhead hunting resumed in 1991 to over 40 – 2 orders of magnitude fewer than the 1,618 bowheads caught by Alaska's Inupiat between 1985 and 2017, according to IWC data discussed by Reeves and Lee.[23]

While Canadian Inuit total bowhead harvests may appear small compared to Alaskan Iñupiat, according to Nobuhiro Kishigami, "the resumption of Inuit whaling is a politically important event in two ways," the first being that Ottawa thereby "publicly demonstrated its political stance concerning aboriginal policy through the revival of the hunt," and the second being that "through the whale hunts the Inuit themselves dramatically embodied their aboriginal rights, in addition to promoting and maintaining their identity."[24] Nobuhiro Kishigami recounts how bowhead hunting by Canadian Inuit had largely disappeared for half a century, but with the settlement of Inuit land claims starting in 1975 and continuing for the next forty years,

> whaling has been considered to be one of their traditional hunting and fishing activities and therefore a right. For that reason, the Canadian government, based on its own judgment, permitted the Inuit to hunt bowhead whales within domestic waters for the purpose of subsistence. As a result, in 1991 the Inuvialuit in the Northwest Territories, in 1996 the Inuit in the Nunavut Territory, and in 2008 the Inuit in northern Quebec Province (Nunavik) all resumed bowhead whaling after long interruptions.[25]

The 1991 Inuvialuit Bowhead Hunt

The Spark That Ignited a Renaissance in Canadian Inuit Bowhead Hunting

Just as their historic comprehensive land claims settlement, the 1984 IFA has enabled substantial Inuvialuit participation in, and regulatory oversight of, the development of the Inuvialuit Settlement Region (ISR)'s natural resource economy and it has concomitantly enabled the Inuvialuit to restore their age-old tradition of subsistence harvesting of bowhead whales, a cornerstone of Inuvialuit traditional culture. For ages, the Inuvialuit lived off the land and waters of the Western Arctic region, harvesting land and marine mammals, including beluga and bowhead whales. Bowhead stocks were severely depleted by the Yankee whaling fleet at the start of the 20th century, and the Inuvialuit had stopped their subsistence bowhead whale hunts during the 1920s, shifting to the more bountiful beluga whales that had not been decimated by the fleets of commercial whalers operating in Arctic waters. Interest in resuming bowhead hunting re-emerged in the 1960s, and while sustained, no hunts would take place until 1991, when, on 3 September 1991, just seven years after the signing of the IFA, the Inuvialuit of the Western Arctic – see Figure 8.1 – successfully, and *legally*, harvested their first bowhead whale in over half a century.[26]

As noted by Lois A. Harwood and Thomas G. Smith,

> [t]here appears to have been little subsistence whaling for the bowhead during or after the commercial whaling era. The last bowhead whale landed by the ancestors of the Inuvialuit in Canada's Western Arctic was taken in 1926 at the Baillie Islands.[27]

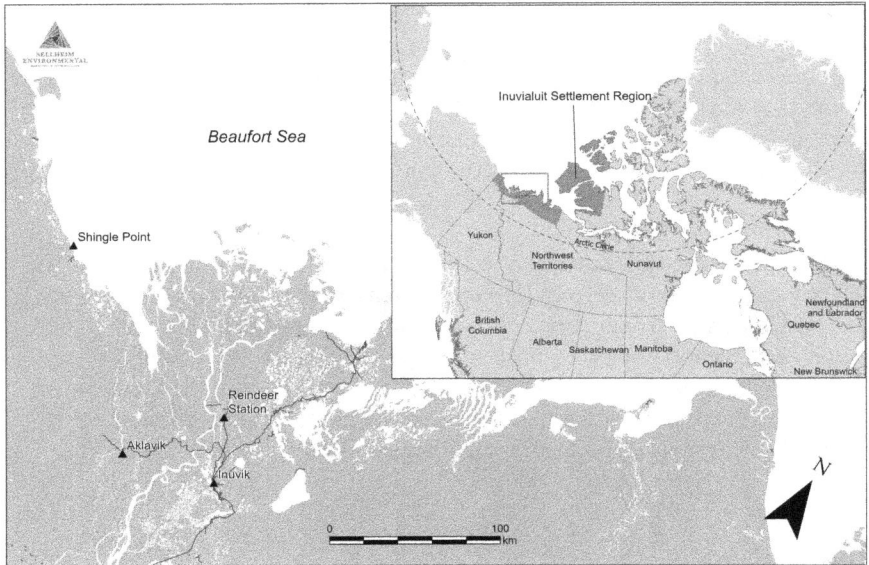

Figure 8.1 The ISR and the northwest Canadian Arctic.

But since the early 1960s, Inuvialuit interest in subsistence bowhead hunting revived, and:

> An internal government report dated 1963 is the first written record of the people's interest in resuming bowhead whale hunting. In 1965, the Department of Northern Affairs and Natural Resources issued licenses to the communities of Aklavik, Tuktoyaktuk, and Sachs Harbour to take one bowhead whale each. No bowheads were harvested under these licenses. Other attempts by residents of Aklavik to resume bowhead hunting took place in 1969, 1975, and 1979, but no hunting took place. In 1988, the Aklavik Hunters and Trappers Committee (HTC) prepared a formal proposal for a license to land one, or strike two, bowhead whales from the waters off Shingle Point, Yukon. After a lengthy consultative and review process, the DFO issued the license to the Aklavik HTC in 1991, and one subadult male bowhead whale (11.1 m long) was landed at Shingle Point on 3 September of that year.[28]

This historic occasion marked the restoration of large cetacean harvesting for subsistence purposes by Inuvialuit and inspired the Inuit of Nunavut and Nunavik to follow in their footsteps after their respective land claims were finalized (in 1993 and 2007) and thus marked a symbolic victory for the Inuit in their historic struggle to restore their rights. The 1991 bowhead harvest was a result of the commitment of the Aklavik Hunters and Trappers Committee, which had been trying to get a bowhead license for several years, the Inuvialuit Game Council, the federal Department of Fisheries and Oceans (DFO), the Inuvialuit Joint Secretariat and the Fisheries

Joint Management Committee (FJMC), which under the terms of Sec. 14(61) of the IFA was created "to assist Canada and the Inuvialuit in administering the rights and obligations relating to fisheries under this Agreement and to assist the Minister of Fisheries and Oceans of Canada in carrying out his responsibilities for the management of fisheries," and to advise the Minister of Fisheries and Oceans "on matters relating to Inuvialuit and the Inuvialuit Settlement Region fisheries."[29] The political climate preceding the historic bowhead harvest was defined by a spirit of cooperation between Inuvialuit and Government; but at the international level, there was much tension, particularly from the United States and the IWC.

That the hunt was approved by the Minister of Fisheries and Oceans was in and of itself a major success for the co-management regimes established by the IFA in 1984. The legal structures established by the IFA were certainly challenged by the conflicting diplomatic demands of the IWC, an international regime regulating large whale harvesting, and by the omnipresent threat of sanctions by Canada's largest trading partner, the United States. With a fisheries conflict simmering in Atlantic Canada and along the Northwest British Columbian coast, the Inuvialuit bowhead harvest brought Canada's external conflict with its neighbour to an unlikely fishery: the western Beaufort Sea near the Mackenzie Delta. In spite of diplomatic complaints by the U.S. State Department, Canada's Minister of Fisheries stood his ground and issued the historic license at the end of August 1991. The rest, of course, was history.[30]

An "Historic" Day

History was made that crisp, clear September day in 1991, when the Inuvialuit of Aklavik restored their age-old tradition by harvesting the first bowhead whale in Canadian waters since the 1920s.[31] The bowhead was first struck at 4:30 p.m. on 3 September 1991: at first, the harpoon gun wouldn't fire, due to a faulty firing pin. The very first strike was delivered by Captain Danny A. Gordon's boat crew, which included, in addition to Gordon, harpooner Titus Allen and Billy Archie. A second back-up boat was present, with Robert Arey, George Allen and Moses Kayotuk onboard. After the first strike, the whale dove to the sea bottom and presumably rolled, removing the harpoon, which bent and disengaged from the whale. For 45 minutes, the whale appeared lost. No float was attached, but the hunters managed to find the whale after following the oil and blood trail. There were another four strikes, but one of the bombs did not go off. Caution and care were the best words to describe the leadership of Captain Danny A. Gordon. DFO sources at Shingle Point said that he held back the hunters until the best whale was found, passing up an opportunity to strike a female that surfaced near the hunters, because its calf was spotted nearby. As well, because of his skilled leadership, the hunters avoided striking a gray whale that surfaced nearby among the bowheads.

Spokespersons with the DFO, the Joint Secretariat and witnesses to the hunt say the whale died within three hours of the first strike. One observer from the Joint Secretariat explained this was relatively fast, noting that in Alaska sometimes a whale might last for 12 hours after being wounded during the initial strike. Rough seas made for a difficult chase, but by closely following the trail of blood and occasional misty sprays through the whale's blowhole, the hunters were able to find and secure the bowhead after the chase. By poking and probing with harpoons, its tail fluke was found. This was tied up, with the crew of hunters and support crew hauling the tow-line up onto the iceberg as the tail was secured. The size of the whale was 36.5 feet (ca. 11 m). It was a young male bowhead. The whale was finally secured by tow ropes at an ice floe by 10:30 p.m. just a few miles west of Shingle Point near Whiteman Hill. With help from two support boats and the two hunting boats, the medium-size bowhead was towed to Shingle Point for six hours through the night, beginning at 11:15 p.m., and being landed at 5:00 a.m. on September 4th. The weather was clear, with an East-Northeast wind blowing at 10–15 miles per hour. The seas were relatively calm with a two to three-foot swell. Surrounded by ice, and beneath the shimmering beauty of the northern lights, the successful hunters revived a traditional that goes back before the time of their grandparents and great grandparents.

Throughout the day visitors from the region joined those already there at Shingle, to help pull the whale up on shore, to help process the *muktuk* – whale skin and blubber – and meat, to help prepare food and hot coffee to keep those involved going as the weather grew colder and the wind picked up. More than 100 people were on the beach, with 60–70 people on 2 block and tackles, with 2 anchors, to pull the whale onto the beach. Historical photos show the uncanny similarity in the bowhead hunts of yesterday and today: the same block and tackle, known as the "deadman," the same community effort pulling the whale to shore, the same hand tools required to cut open the whale, remove the *muktuk* and meat and separate the heavy head from the body prior to hauling it ashore. Boats streamed down the Mackenzie River to the coast, and chartered aircraft of all sizes found their way to Shingle Point. Friends, family and local media joined the hunters and their supporters at Shingle Point for a day of excitement, preparing the *muktuk*, removing meat and blubber from the whale, and enjoying freshly boiled bowhead *muktuk*, hot coffee and bannock.

By the day's end, some 200 people were congregating at Shingle Point, as the whale was butchered, its meat and *muktuk* cut and dried on the pebbly beach. Just as their forefathers harvested bowheads, using the combined strength of the whole community to haul the whale out of the sea, and to prepare the meat and *muktuk*, the Inuvialuit worked together all day long and into the next day. The first cut of blubber was made at around 12:30 p.m., and one third of the *muktuk* was removed by 4:10 p.m. The first cut of meat was made at 4:30 p.m., and the stomach cavity was cut open at 5:10 p.m. By 10:00 p.m., nearly half the whale had been cut up and laid out on the beach. It was expected that up to 250 pieces of *muktuk*, in one cubic foot blocks, and 150 pieces of meat, in the same dimensions, would be available for distribution to the 6 Inuvialuit communities. A traditional drum dance was performed on the beach as the late summer's sun set into the night sky, celebrating the successful restoration of a proud tradition. The hunters and crew got up and danced before the crowd as their forefathers had. A feast and dance were planned for September 27th in Aklavik to further celebrate this historic event. By the end of the week, all the *muktuk* and meat had been shipped up to Aklavik, and the hunters and their crew, and all their supporters who were with them at Shingle, finally had a chance to take a well-deserved rest. Over the coming weeks, bowhead meat was distributed throughout the six Inuvialuit communities by boat and plane.[32]

The historic Inuvialuit bowhead harvest restored a proud tradition: the magic felt on the beach at Shingle Point marked a turning point for the Inuvialuit, with Inuvialuit culture and traditions now on the mend, and the Inuvialuit future now looking bright and hopeful. The excitement of the harvest was felt by all, as the Inuvialuit reached back and revived a tradition in harmony with the landscape of the Western Arctic. Years of hard work, beginning with Aklavik resident Dodie Malegana's first application to the DFO for a license to hunt a bowhead a few years back, and culminating with the month-long hunt at Shingle, paid off. As well, the lasting legacy of the IFA, which guaranteed the Inuvialuit right to harvest marine mammals as they have always done, and constitutionally protects this right, was demonstrated in an eloquent and meaningful way, showing how the hunters and trappers from the six Inuvialuit communities can benefit from their land claim, and how co-management can work to protect their traditions, even in the face of stiff resistance by outsiders.

Co-Management in Action

The restoration of subsistence bowhead hunting by the Inuvialuit in 1991 demonstrates how co-management can succeed in protecting subsistence and helping to revitalize an ancient tradition using a modern decision-making structure.[33] Though the final decision for the bowhead license rested with the Minister of Fishers and Oceans, a rejection would require not only a written explanation with 30 days to the FJMC, but there would also be an opportunity to respond to the Minister's written explanation to which the Minister would again have to reply in writing within 30 days, putting the final decision under a microscope for all the world to see.[34] So while the Minister was no doubt under other pressures, particularly the threats of trade sanctions by the United States, and to a lesser degree, the opposition by the IWC and other environmental groups, with large constituencies in the South, he would remain accountable to the Inuvialuit, and on the record. For him to say no, he would have to be ready to defend his rejection and face the consequences. And in those months following the armed Mohawk uprising at Oka,[35] Ottawa was not inclined to create any further provocations or suggest anything less than a whole-hearted commitment to Aboriginal rights.

A look at the Inuvialuit bowhead license decision presents an excellent opportunity to apply structural theories of international relations, and in particular the levels-of-analysis – such as Kenneth Waltz's classic three images or J. David Singer's original two levels of analysis, or this chapter's own author's revision of the Waltzian framework with the addition of a "fourth image" that restores tribal polities, located between local governments and national governments – to the ecosystem of local, regional, national and international structures that emerged in the post-land claims settlement era of the Western Arctic. Indeed, there appear to be three salient levels of analysis useful for understanding the significance and dynamic of the 1991 Inuvialuit bowhead harvest.

The first may be thought of as a tribal level of analysis (which I dub the "fourth image"), though it falls between the individual ("image one") and the national-state ("image two"), representing indigenous peoples and their polities at the regional sub-system level, such as that established in 1984 by the IFA, which created the ISR representing the now formally recognized indigenous boundaries of the Inuvialuit homeland as a distinct and increasingly autonomous region. Inside the ISR, resources are now co-managed under the principles encoded by the IFA, and the Inuvialuit retain rights to subsistence harvesting of all wildlife, to be managed by co-management boards like the FJMC and the Wildlife Management Advisory Councils. The second level of analysis was the national level ("image two" in the Waltzian framework), with the DFO serving in a lead role. And the third was the international ("image three" in the Waltzian framework), with a bipolar regional structure marked by economic interdependence, strategic partnership and diplomatic competition to moderate economic disagreements. The third level of analysis harbors numerous international organizations that manage a variety of global and regional issues, such as the IWC, as well as activist entities such as Greenpeace and Sea Shepherd; due to Ottawa's withdrawal, the IWC was left largely on the sidelines, and neither Greenpeace nor Sea Shepherd showed up at Shingle Point or lobbied actively to thwart the Inuvialuit bowhead hunt, despite Inuvialuit worries this could transpire.

At the third image, tensions caused by international trade had precipitated an ongoing, low-intensity political conflict where coercive diplomacy, economic sanctions, tariffs and import duties were all part of the diplomatic landscape, as was a close strategic partnership as represented by their joint North American Aerospace Defense Command (NORAD) membership. It was not uncommon, therefore, for unrelated issues to be brought together by policy linkage at the national level. The potential for such sanctions is omnipresent, subject to the whim of the overall relationship between Canada and the United States; as a result, to overcome such threats by such an important trade and strategic partner as the United States required substantial political will in Ottawa. This was clearly demonstrated by the federal government's commitment to endorse the Inuvialuit application for a bowhead license in 1991.

Issuing a license to harvest a bowhead whale in Canadian waters, while in opposition to the IWC's protests (anaemic as they were in this case), was fully consistent with both the IFA and the Fisheries Act, and thereby entirely legal in Canada – and because Canada was by 1991 just an observer to the IWC, and because authorization for the hunt was made at the national and local levels and outside the IWC's jurisdiction, the only real sanction that the IWC could muster to demand compliance, particularly from an observer nation and not a full member, would be through diplomatic protest or some sort of trade sanction imposed by like-minded member states, most likely led by the United States.

However, the moral high-ground once occupied by the IWC had been greatly eroded, largely by its unwillingness to accept whaling where stocks could sustain a harvest, commercial or subsistence, and by its insensitivity to Aboriginal peoples and their constitutionally recognized rights to engage in subsistence harvesting, enshrined in the 1982 Constitution Act and spelled out in detail in the 1984 IFA. This reduced the likelihood of sanctions being imposed by the United States on Canada for its authorization of the Inuvialuit bowhead harvest, given the sacrosanct constitutional and historical issues at play in Canada. Ironically, the ability of an international organization like the IWC to impose its will boils down to domestic pressures and the ability of constituent member states to impose sanctions; and this, more or less, boils down to the ability (and willingness) of the United States to do so given its enormous trade relationship with Canada. That the Inupiat of Alaska's North Slope harvest dozens of bowheads each year, with a landed quota at the time of 41 per annum, and that they share with Inuvialuit a common historical, cultural and linguistic tradition, reduced the likelihood of any such punitive measures, which indeed never came to pass despite palpable anxiety in the Western Arctic at the time that they might.

Celebrating the Rebirth of a Proud Tradition

More than 350 visitors crowded the Sittichinli community centre in Aklavik on 27 September 2001, where one of the most festive celebrations in recent history was held. People came from the coastal and delta communities, bringing with them slippers for jigging and square dancing.[36] There was drum dancing, and speeches were delivered by the bowhead hunters. Harpooner Titus Allen recalled the wishes of the elders, many of whom had passed on before their dream to hunt the great bowhead whale again was realized by the hunters of Aklavik. Many letters of congratulations and thanks from the community leaders, students learning about their traditions and many of those unable to attend the festivities were read to an appreciating crowd.

As Grade Six student Jonathan Ettie wrote: "To the hunters of the bowhead whale, I'm glad you caught that whale. It must have been an exciting experience! I wish I was with you." Peter Hansen, also a Grade Six student from Inuvik, wrote:

Dear Danny A., Hi. I am from Inuvik and I wanted to tell you the bowhead *muktuk* was pretty tasty. You made a good catch on that one. And I would like to thank you for letting us sample some bowhead. I have never tried it before.

Patrick Lancaster wrote:

Dear Hunters, I'm really glad you caught that whale. There must have been plenty of meat to go around. When our teacher heard about it, she made us study whales. It was a very good learning experience. I do have a few questions, though. Do you use all of the whale? How do you eat the *muktuk*? Please send your answers to Patrick Lancaster.

Robert Kuptana, chairman of the Holman Community Corporation and a future chair of the IRC, wrote the following:

> To: Captain Danny A. Gordon, your group the Aklavik HTC, and to the Inuit of Aklavik; We, of the community of Holman, would like to express our gratitude and congratulations to you all on a successful bowhead whale hunt, on behalf of all Inuit. Even though our society in the North is changing fast, the successful bowhead whale hunt you have made will contribute in preserving Inuit culture and identity to present and future values. For that reason, Inuit sense the moral victory and satisfaction that our ancestors were able to practice. Once again, congratulations, a satisfaction well deserved. Thanks be to God.

Throughout the feast, Aklavik residents distributed bowhead *muktuk*, tongue and meat to some very appreciating guests, along with endless amounts of sumptuous traditional food. Some of the food servers wore T-shirts commemorating "The Great Bowhead Hunt of 1991," and "Shingle Point: September 3, 1991." A large cake was presented to Captain Danny A. Gordon, Harpooner Titus Allen and Billy Archie, with a bowhead whale pictured on the frosting. All the hunters were introduced, each receiving an enthusiastic round of applause. An Aklavik teacher and the Grade Two and Three students sang a song called "Once there was a Bowhead" and later on, the hunters joined in the drum dance as their forefathers did before them. The night was filled with old-time dancing, great fiddling and the finest jigging. Tears of joy and hope were mixed with some tears of sadness for those elders who were no longer with them, as the great glory of the Inuvialuit bowhead harvest was celebrated with passion and excitement. As thenNorthwest Territories (NWT) Premier Nellie Cournoyea – who soon thereafter returned to Inuvik from Yellowknife to serve for many years as chair of the Inuvialuit Regional Council (IRC), itself a powerful reminder of the ISR's emergence as a self-governing region deserving of widely respected Cournoyea's leadership as Premier of the NWT – wrote to the people of Aklavik:

> It is always a good feeling to sit down to a feast of good country food obtained from a successful hunt. How much better it must feel for all of you to be at a feast where the menu includes the first bowhead whale harvested since the 1920s. I wish I could be with you to share the food, friends and drum dance. While you are biting into this special feast of bowhead, it is a good time to be thankful that everyone returned from the hunt with no injuries or accidents. This is a hunt with a high level of risk involved. I would also like to say to all the organizers, hunters, watchers, cooks, haulers, sailors, cutters and all who made this hunt possible, congratulations on a successful hunt and well earned feast.[37]

Applying the Canadian Experience to Japanese Whaling: Fostering Multilevel Unity, from the National to the Local Level

As Japan refocuses on coastal whaling in wake of its own more recent 2019 IWC withdrawal, the Japanese whaling industry and its supporters across Japan seek to

spark renewed interest in the consumption of whale meat in Japan, through all manner of innovations from the 2022 deployment of whale-meat food trucks (called
"Kitchen Cars" in Japan[38]), to the installation of whale meat vending machines in
January 2023[39] to pro-active courtship of international social media influencers
two months later through whale-tasting events designed to foster the emergence of
a tourist market interested in experiencing whale cuisine[40] – similar to that which
has arisen somewhat paradoxically side-by-side with the whale-watching industry
in Iceland, and it is thus hoped to reverse or offset through budding tourist interest
the long decline in whale meat consumption by Japanese.

As these efforts gain steam, the experience of Canada's indigenous coastal
whalers, in partnership with the federal government in Ottawa which has steadfastly supported Inuit whaling efforts, offers much insight and inspiration thus mer-
iting our above look-back to the first bowhead hunt in the Canadian Arctic after
its IWC exit: the 1991 bowhead hunt by the Inuvialuit of Aklavik, which I had
the great privilege of covering as editor of the Inuvialuit bilingual Inuvialuktun-
English newspaper, *Tusaayaksat*. There has been notable attention in the literature
on the recent renaissance in indigenous whaling in Canada ever since this back-
to-back dual convergence of Ottawa's repatriation of the Canadian Constitution in
April 1982 and subsequent withdrawal from the IWC two months later, even if the
literature does not explicitly link Ottawa's IWC exit decision to the constitutional
repatriation that preceded the formal date of its IWC withdrawal.

The persistent and growing community interest in restoring large cetacean harvesting, the sustained commitment to these efforts by federal government officials
in Ottawa in partnership with local and territorial leaders in the Arctic, and the
expanding successful track record of bowhead hunting in the Canadian Arctic have
established a strong connection and durable political-economic alignment of powerful actors at the federal, regional and local level that we can expect to be mirrored
in Japan so long as the government in Tokyo continues to support its whaling communities and industry in a concerted effort to rekindle interest in Japan's whaling
heritage, and in contemporary consumption of whale meat, whether by Japanese
residents or visitors to Japan. Indeed, restoring a tradition from the brink of cultural oblivion across a vast land is shown in the restoration of bowhead whaling by
Canadian Inuit to be possible, but not immediate; indeed, it's been a generational
process, one of deep symbolic and cultural importance – and this is an important
lesson to bear in mind as Japan continues its own restoration of coastal whaling,
that a sustained and incremental effort can bear fruit if supported by unity of purpose uniting not just government and people, but the many levels of government
from the municipality to the state.

As Niall Alexander Rand has observed, the IWC "conceals within its history
a perennial battle between nations," one that has achieved something of a stalemate: "Since the moratorium on commercial whaling took effect in 1986 both sides
of the whaling debate have been unable to substantively advance their cause."[41]
Rand turns his attention to the "place of Indigenous peoples within the history of
whaling and what role, if any, they will play in the future relevance of the IWC,"
arguing "that Canada's withdrawal from the IWC, in the interest of its Indigenous
peoples, should generally be regarded as a domestic regulatory success."[42] When

Prime Minister Pierre Trudeau announced in 1981 that Canada would soon exit the IWC, as noted above, he explained that Canada no longer perceived itself to be a whaling nation.

And yet, in the four decades since then, it has become increasingly just that, albeit a nation that fosters indigenous *subsistence* whaling and not *commercial* whaling, with indigenous whaling and its rise across the Arctic perceived by federal, provincial/territorial and local actors as strengthening the very constitutional fabric of Canadian Arctic sovereignty, and with each new bowhead hunt of the approximately 40 bowheads taken in the three decades since the resumption of bowhead hunting by Canadian Inuit, this gathering restoration of Inuit whaling culture reaffirms Ottawa's governance of its vast Arctic region through its enduring support of and partnership with the Inuit, whose long presence in and continued residency of the Arctic helps offsets Ottawa's otherwise tenuous assertion of sovereignty over its vast northern lands and coastal waters.

Prime Minister Pierre Trudeau noted the resistance Ottawa faced from environmental activists who viewed its IWC membership as *de facto* approval of whaling, inclusive of commercial whaling, an economic activity Canada no longer partook in. The re-emergence of a vibrant and passionate indigenous subsistence whaling culture in the Canadian Arctic thus introduced a demonstrably more sustainable approach to whaling, one that eventually has won approval of some of the most vocal anti-whaling organizations, such as Greenpeace, which has come to gradually accept Inuit rights to subsistence hunting, despite its own historical animosity to Inuit hunting traditions, whether of whales or seals, and issued an apology to the Inuit in 2014 for its anti-sealing campaign that decimated the economies of remote Inuit villages, now pledging: "For our part, we'll do whatever we can to […] acknowledge the power of Indigenous knowledge, and seek to apply the lessons of their holistic approach to a new model of conservation."[43] The coalition uniting multiple levels of government, from the federal to the provincial/territorial to the local level, in a joint effort to restore Inuit cultural traditions and affirm the constitutional protection of indigenous rights enshrined in the 1982 Constitution Act, has thus extended beyond government to include another sector of the populace: the largely urban environmental activist community, joining the remote Inuit populace to form a broad and enduring coalition of both government and people.

One may expect something comparable to this to take root in Japan as well, albeit within the Japanese constitutional, political and cultural context, much as we have also seen in Iceland where coastal whaling within that island state's EEZ is now widely perceived as legally protected assertion of sovereignty and legitimate implementation of the rights enshrined by United Nations Convention on the Law of the Sea for nations to reap the bounty of their own territorial waters, whether for subsistence or commercial purposes, whose preservation is perceived to complement the very exercise of national sovereignty itself; without its prior vast global scale and thus in marked contrast to the international whale trade of the 19th century, which so devastated whale stocks around the world, coastal whaling reinforces, and thereby strengthens, national unity and stabilizes projections of sovereignty in remote coastal regions.

Coastal whaling in Japan, as with renewed Canadian and Icelandic whaling, efforts in recent decades, in essence *right-sizes* whaling from global to national and local scales, bringing it within a domestic constitutional and economic framework and embedding it in a tightly controlled management regime to ensure quotas remain sustainable, a pre-requisite for extending a whole of government approach to revitalizing whaling culture to a broad-based popular coalition that includes urban (and often environmentalist) communities in addition to local whaling communities.[44] Indeed, this aligns well with the observations of Dennis Normile, in the pages of *Science*:

Japan's 26 December 2018 announcement that it will withdraw from the International Whaling Commission (IWC) and resume commercial whaling in its own waters triggered fierce criticism around the world. U.K. environment secretary Michael Gove was 'extremely disappointed.' Greenpeace called the decision 'out of step with the international community' and its timing in the middle of the holiday season 'sneaky.' But some conservationists say the hand wringers are missing the point. What matters most is that Japan has decided 'to stop large-scale whaling' on the high seas under the mantle of scientific research, says Justin Cooke, a marine population assessment specialist at the Center for Ecosystem Management Studies in Emmendingen, Germany. Given the declining appetite for whale meat, Japan is unlikely to start to catch many more whales in its own waters than it already does, he adds: 'There won't be much change on the ground.' Patrick Ramage, a whaling specialist at the International Fund for Animal Welfare in Yarmouth, Massachusetts, agrees. 'It's good news for whales,' he says – and also for IWC, which can finally end its 'food fights over whaling' and focus on other issues in whale conservation.[45]

Indeed, sustainability is front and centre to Japan's pivot from high seas scientific whaling to internal coastal whaling. As described by the director of Japan's Ship and Ocean Foundation (SOF), Hiroshi Terashima, "As mankind moves into the 21st century, integrated policies of ocean governance are necessary for the sustainable development and use of our oceans and their resources and the protection of the marine environment," and in support of this, SOF "has launched an 'Institute for Ocean Policy,' with the mission statement 'Living in Harmony with the Oceans'."[46] Sustainable efforts to live in harmony with the oceans extend to Japan's restoration of coastal whaling since 2019, which are aimed specifically at harvesting whales for their meat, part of a long cultural tradition in this maritime nation dating back millennia just as is the case in the Canadian Arctic.

As Joji Morishita describes in the same SOF newsletter:

One thing must first be made clear about the discussion on whaling: the Government of Japan aims at the sustainable use of only abundant species of whales [...] while protecting depleted whale species consistent with international law including the United Nations Convention of the Law of the Sea

(UNCLOS) and the International Convention for the Regulation of Whaling (ICRW). People who are against whaling claim that whaling in the past depleted whale resources and therefore sustainable whaling is impossible in the future. The fact is that whale resources were overexploited in the past by whaling for whale oil which was used for industrial purposes. In that sense, whaling for food has become a victim of the past overexploitation. [...] Because the current demand for whales as food is far smaller than the past demand for whale oil, it can be easily imagined that the overexploitation experienced in the past will not happen again.[47]

Morishita further explains that:

with greatly advanced scientific knowledge about whales and wildlife management and new technologies available today, sustainable whaling is possible. Whaling was conducted in the past with insufficient knowledge and without using the technologies that we have today: a scientific and risk-averse method for calculating catch quotas that specifically accounts for uncertainty to ensure that utilization of abundant whale resources can occur without depletion, a device for monitoring positions and movements of whaling vessels with satellite technology, and a technique for identifying each individual whale based on DNA analysis to ensure that management rules are followed.[48]

With whaling increasingly aligned with values of sustainability, we see not only a convergence between subsistence and commercial whaling but also a convergence in the rights of whalers and the rights of whales, long a dividing point that pitted whalers and animal rights activists when they were in many ways natural allies. The running skirmishes in the Southern Ocean between Sea Shepherd and Japanese whalers are now firmly rooted in the past, as sustainable coastal whaling finds its stride. As we see in Arctic Canada, the old tensions between Inuit whalers and animal rights activists are likewise an artefact of an earlier time, as new alignments of interest bring groups like Greenpeace closer to Indigenous communities and organizations, particularly their joint commitment to sustainability. With sustainability at the forefront of whaling practices, the rights of whalers and the rights of whales are now more closely aligned, allowing for a renaissance in both indigenous and animal rights fostered by this new alignment. Something parallel to this can be anticipated to emerge in Japan and may in fact already have done so as Sea Shepherd and Greenpeace turn their attention to other struggles and bring a close to their earlier campaigns to save the whales from whalers, now that they can work together with coastal whalers to save whale ecosystems and ensure harvests are sustainable – protecting the rights of both.

In addition to their shared commitment to sustainable whaling with the Inuit of the Canadian Arctic, and their shared realization that achievement of an historic restoration of sustainable whaling inside their respective EEZ required withdrawal

from the IWC, Japanese whaling shares with Canadian Arctic whaling its long tradition as part of the cultural heritage, and while once vibrant, like in the Canadian
Arctic, entered into a long decline and nearly died out; this decline does not reduce
its cultural significance but necessitates a multilevel, whole-of-government effort
to revitalize it. As Morishita describes:

> Even though most Japanese people do not wear Japanese kimono or go to
> watch Noh plays every day, they are undoubtedly part of Japanese culture.
> The contention that whale meat eating is not a part of Japanese culture since
> not all Japanese people eat the meat in everyday life, therefore, does not
> stand.[49]

This parallels much the way Inuit bowhead hunting in Canada restored a long-
enduring cultural tradition of coastal whaling for subsistence purposes by Canadian Inuit, helping Canada to become a whaling nation once again but this time a
sustainable whaling nation committed to the preservation of indigenous traditions,
values and rights within a shared commitment to sustainability that extends from
the Inuit village-village all the way to the national level in Ottawa, protected firmly
by the Canadian Constitution since its 1982 repatriation.

Thus, while exiting the IWC, Canada renewed its commitment to indigenous
whaling, comparable in both scale and heritage to today's sustainable coastal whaling in Japan in support of the nation's centuries-long and proud whaling tradition.
While the two departures from the IWC – Canada's in 1982 and Japan's in 2019 –
are separated by four decades, they share many commonalities and both align
with a broader trend towards a decline in international-scale whaling in favour of
community-scale coastal whaling, a domestic, indeed broadly defined indigenous,
pursuit in both nations, that dates back centuries and in many cases millennia.
While the 2019 exit by Japan from the IWC is still very recent, a look back on
the resumption of large cetacean harvesting by Canadian Inuit four decades ago,
and the first Canadian Inuit bowhead hunt in modern times by the Inuvialuit of
Canada's Western Arctic (connected by kin and culture to the Iñupiat whalers of the
Alaska North Slope) in 1991, less than one decade after Ottawa's 1982 IWC exit
and prior constitutional repatriation, offers us an opportunity for understanding and
anticipation of what we can expect in the years ahead as whaling in Japan ceases to
be a global activity pursued in the Southern Ocean and instead becomes a national
activity within Japan's own EEZ governed by principles of sustainable quotas.

While the resumption of Inuvialuit bowhead hunting was a moment of great
pride and cultural renewal and of much symbolic importance, after the historic
1991 bowhead hunt there followed only one more hunt in 1996. Further east, the
Inuit of Nunavut have in turn resumed bowhead hunting, particularly in the years
that followed the formation of the Nunavut Territory in 1999 – and Inuit of Nunavik were awarded a quote in 2008, a year after settling their maritime land claim
agreement. Despite the increase in Inuit communities hunting bowheads in both
Nunavut and Nunavik, collectively they have hunted only around 1 or 2 bowheads

per annum in the years since the Inuvialuit became the first Canadian Inuit to resume bowhead hunting in 1991, with little over 40 bowheads killed in total since
Ottawa's 1982 IWC exit and constitutional repatriation. As Kishigami writes:

> From 1991 to 2008 hunts were carried out about once every two years, and
> the number of whales caught (the quota) gradually increased. Each year since
> 2008 the Canadian government has allowed three whales to be caught in the
> Nunavut Territory, one in Nunavik, and one in the Northwest Territories.[50]

As in coastal Japan, Canada's northern peoples have a long, proud tradition of
coastal whaling for subsistence purposes. The technologies used, the risks and dan-
gers overcome, and the importance of whale meat to the survival of entire communities are common across these two maritime nations. Each time a whale is
caught by Canadian Inuit, it instills a deep sense of pride and passion throughout
the community. What started as one bowhead harvested in 1991 grew to two in
1996 and soon thereafter has become a largely continuous activity, each time reaf-
firming Inuit values and traditions and celebrating the restoration of Inuit rights.
It is no coincidence that famed anthropologist of Inuit whaling Milton Freeman
entitled his 1992 book on the Inuvialuit bowhead hunt *Recovering Rights*, with the
double-entendre intended to convey both Inuit cultural rights and whale stocks of
the bowhead whale, known to whalers as "right whales" for their tendency to float
after capture and not sink to the bottom of the sea.[51]

As reported in *Nunatsiaq News* a decade ago, more than two decades after the
Inuvialuit restored the tradition of Inuit bowhead hunting to Canada and yet captur-
ing the very same sense of community pride and excitement:

> Repulse Bay is ready to celebrate a successful bowhead whale hunt. A hunting
> party from the community finally caught a whale measuring 51 feet and seven
> inches long (15.72 meters) on Aug. 31, after 10 days of repeated efforts in shifty
> weather. [...] This year's bowhead whale hunt is Repulse Bay's fifth since 1996,
> when the Fisheries and Oceans minister first approved the harvest of one bow-
> head for the Nunavut Settlement Area. That quota was increased to three bow-
> head whales in 2009, with three Nunavut communities, one per region, receiving
> permits to hunt single bowhead whales since then. Repulse Bay also harvested
> bowheads in 2005, 2010 and 2012. Their biggest catch was in 2010 when the
> community harvested a bowhead whale almost 53 feet and 10 inches long.[52]

Fast-forward another eight years, and Inuit bowhead hunting continues to thrive in
Arctic Canada, as described in the Nunavut community of Baker Lake, a neighbour
to Repulse Bay (now known by its traditional Inuit name, Naujaat):

> The hunters brought their catch to shore on Tuesday and began harvesting
> the maktaaq. They shared it with people in Naujaat, and people from that
> community helped cut up the whale and packaged the rest of the maktaaq for
> every community in the region. 'The whole town came out and celebrated,'

he said. 'There were shouts and cheering and a song was sung.' Jamie Kataluk is a member of the team ... When talking about the experience, he was overwhelmed. 'The Inuit way of life – people gathering together to achieve one goal. And to see that happening here, it was, it was – sometimes it's hard to describe,' he said.[53]

And it is very difficult to capture in words what is felt so clearly and strongly, much emotional power is successfully conveyed, just as it has since the Inuvialuit landed the first legally caught bowhead whale in Canadian waters in over half a century back in 1991.

A similar process is now under way in Japan, as coastal whalers, their communities and governments from the municipal to the national align to bring whaling back as a coastal and national activity, free from the struggles on the high seas that were associated with its scientific whaling program that reached from the North Pacific to the Southern Ocean. New, modern methods of distributing whale meat, from vending machine to food carts, and to cultivate contemporary markets with a taste for consuming it, have been introduced in recent times, in the hope the declining popular appetite for whale meat in Japan can be reversed – and the opportunity presents itself for whaling in Japan to experience the very same sort of renaissance that it has enjoyed in the Canadian Arctic, should not only the stars align, but with the stars the political will and support of governments of every level, and communities as diverse as remote fishing villages and modern urban megalopolises. It won't necessarily be easy, or quick, nor will it happen without a broad coalition of stakeholder support. And it's not a given that such a coalition will align.

Consider the observations in the whaling community of Wada by *The Guardian's* Justin McCurry:

In 2019, when Japan withdrew from the International Whaling Commission (IWC) – the body that had effectively banned whaling in the late 1980s – Wada rejoiced at the prospect of a return to commercial hunting and at a popular reconnection with a source of food that had sustained coastal communities for 400 years. But here and in other whaling towns in Japan,[54] the resumption of killing whales for profit for the first time in more than three decades has offered little cause for celebration. While condemnation from conservation groups has eased in the three years since Japan's fleet exited the Antarctic, the country's whalers face other obstacles: ageing fishermen and vessels, mysterious changes in cetacean behaviour possibly linked to climate change, and a stubborn refusal among Japanese people to eat enough whale meat to make killing them a profitable venture.[55]

Indeed, the numbers do appear worrisome, as:

Barely 300 people in Japan are directly connected to whaling, while whale made up only about 0.1% of the country's total meat consumption in 2016, according to government data. About 4–5,000 tonnes of whale meat enter the domestic market every year – the equivalent in volume of about half an apple for every person.[56]

But the small size of the whaling community does not negate its cultural importance, and may be more a reflection of the multiple vectors of modernization in society that has similarly threatened the viability of various components of Japanese cultural heritage and traditions, as Morishita above noted with regard to the declining numbers of Japanese who wear traditional *kimonos* or attend *Noh* plays; small numbers do not convey reduced cultural significance.

Indeed, the renaissance in Inuit bowhead hunting involves a similar order of magnitude of participants, and when compared to Canada's overall population, it would appear by the same sort of metric to convey the same sort of fragility. But keepers of traditional knowledge have, in the modern world, often been a small but no less important minority. And, when supported by their communities and governments, it can help to keep traditions alive and to find a new generation of practitioners. This indeed align well with what Yoshinori Shoji, president of the Wada-based whaling company Gaibo Hogei, told McCurry, that "abandoning coastal hunting was unthinkable. 'I know it is controversial in other parts of the world, but for us, whales are simply a source of food,' said Shoji, whose company has been processing whale meat for more than 70 years."[57] In order to continue to do so for the next 70 years will require continued support from the Japanese government, including its generous subsidy to the whaling industry. Otherwise, the high cost of whaling, and the present-day disinterest among most Japanese with regard to consuming whale meat, could become an insurmountable challenge.

As Japanese journalist Junko Sakuma explained to *The Guardian*'s McCurry.

Japan's commercial whaling industry would grind to a halt without government subsidies of ¥5.1bn (£.033bn) a year. […] The government has said that it can't continue to subside what is supposed to be a commercial concern for ever. When Japan left the IWC, fisheries officials thought they would be able to catch as many whales as they needed to sustain the industry, but in fact it has shrunk. Japanese whaling will continue, but in a much smaller form.[58]

Ironically, some observers note that Japan's pivot from scientific whaling on the high seas, with all its controversy and dramatic confrontations with Sea Shepherd, to coastal whaling within Japan's EEZ, has contributed to the decline in interest among Japanese, who responded to international criticisms of Japanese cultural heritage with an impassioned defence of those traditions. As McCurry reported in *The Guardian*,

Paradoxically, the end of 'scientific' whaling and the Japanese fleet's annual clashes with the anti-whaling organisation Sea Shepherd may be hastening whaling's decline. 'In the past, Japanese people were defensive because they didn't like white people telling them not to eat whale meat,' Sakuma says. 'But whaling is barely mentioned these days by anti-whaling countries like

Australia, Britain and the US. Now Japanese people have nothing to rebel against, so they could end up just forgetting about whale meat.'[59]

But a "much smaller form" is precisely what Canadian Inuit whalers have embraced. While bowhead hunting has since the "Great Bowhead Hunt of 1991" become an ongoing re-assertion of traditional Inuit culture that has now endured more than a quarter century, the high cost of conducting a bowhead hunt has limited the frequency with which individual communities undertake such hunts. The Inuvialuit, who were the first to receive a license from Canada's Department of Fisheries to hunt a bowhead, waited five years before their second hunt and since then have not yet pursued their third. But interest in resuming bowhead hunting has spread eastward, from the ISR to the Nunavut Territory, and from there to Nunavik in northern Quebec, catalysed by their respective land claim and self-government agreements. While dozens of communities, and cumulatively, many hundreds of Inuit residents of those communities, have participated in bowhead hunting since 1991, each year, the number of people involved in the practice of bowhead hunting can be counted in the dozens or hundreds, and not thousands. It is the undiminished spirit, and not the magnitude of participants, that has kept the bowhead hunting tradition alive, with financial and political support from multiple stakeholders from the local to the national level. The restoration of Inuit bowhead hunting is a renaissance of small but persistent numbers, symbolically powerful but not economically self-sufficient and certainly not commercially feasible.

Whether such a renaissance can be successfully nurtured in Japan remains to be seen, but with the fifth season of coastal whaling under way, there is every hope that as the years become decades, history will repeat itself, and past, as exemplified by the Inuvialuit bowhead hunt of 1991 – catalysed in part by the Canadian IWC exit of 1982, and in turn precipitating dozens of bowhead hunts in the four decades since across the vast length and breadth of the Canadian Arctic – can become prologue for Japan's current effort to renew its long, proud tradition of coastal whaling, and the economy and culture to sustain it, as the Canadian Inuit have achieved.

Notes

1 Oftentimes serving the same clientele, which seeks to witness the majesty of whales in nature and then seeks to taste whale meat once back onshore – paradoxical to some, but upon a closer second look, more compatible than polarized politics might initially suggest.
2 Particularly evident at the IWC, where anti-whaling sentiment has long frustrated whaling nations seeking to continue whaling in contemporary times
3 Where it faced off against piratical NGOs such as the Sea Shepherd Conservation Society (a.k.a. 'Sea Shepherd') and Greenpeace in a recurring series of Antarctic high-sea skirmishes and dangerous rammings.
4 No longer empowered by ice-hardened pirate vessels and thus more an irritant than existential threat to the industry.
5 *New York Times,* 'AROUND THE WORLD; Canadians withdraw from whaling commission,' (New York 28 June 1981) 4.

6 Andrew P. Hutton, 'Canada's intention to withdraw its membership from the International Whaling Commission,' 29 June 1981, *UPI Archives*, <https://www.upi.com/Archives/1981/06/29/Canadas-intention-to-withdraw-its-membership-from-the-International/5126362635200/>.

7 Ibid.

8 *New York Times*, 'AROUND'.

9 See Constitution Act of 1982, https://caid.ca/ConstAct010208.pdf.

10 Ibid.

11 Ibid. In particular, see Section 35 (3) in Part II of the Constitution Act of 1982 on the constitutional empowerment of land claims agreements as a vehicle for the restoration and protection of indigenous rights.

12 Robert Sheppard, 'Reality check: Leviathan under siege. Why is Canada silent as commercial whaling takes off?,' *CBC News in Depth* (Ottawa, 1 May 2006) <https://www.cbc.ca/news2/background/realitycheck/sheppard/20060501.html> accessed 23 April 2023.

13 Now Inuit Circumpolar Council.

14 Randall R. Reeves and David S. Lee, 'Bowhead whales and whaling in the central and eastern Canadian arctic, 1970–2021', (2022), 23 *IWCJCRM*, 7.

15 Ibid., 7.

16 Ibid., 7.

17 Ibid.,7.

18 Ibid., 7.

19 Ibid., 8.

20 Ibid., 8.

21 Ibid., 16.

22 Ibid., 16.

23 Ibid., 8–9.

24 Nobuhiro Kishigami, 'Research note: Revival of inuit bowhead hunts in arctic Canada,' 16 JRCA, 43.

25 Ibid., 48.

26 The details presented in this section on the historic 1991 bowhead hunt were originally reported by the author in *Tusaayaksat* from 12 August 1991, through 7 October 1991, and in 'The great bowhead hunt of 1991', in Barry Zellen (ed.), *The inuvialuit bowhead harvest of 1991: A pictorial history and analysis* (Inuvialuit Communications Society 1992), 33.

27 Lois A. Harwood and Thomas G. Smith, 'Whales of the inuvialuit settlement region in Canada's western arctic: An overview and outlook', (2002), *Arctic* 50(SUPP.1), 80.

28 Ibid., 82.

29 *Inuvialuit Final Agreement (IFA)*, 1984, Sec. 14(61), https://irc.inuvialuit.com/sites/default/files/IFA.pdf

30 For a detailed look at the revival of Inuvialuit bowhead harvesting, see Milton M. R. Freeman, Eleanor E. Wein, and Darren E. Keith, *Recovering rights: Bowhead whales and inuvialuit subsistence in the western Canadian arctic* (Canadian Circumpolar Institute 1992). Also see Barry Zellen (ed.), *The inuvialuit bowhead harvest of 1991: A pictorial history and analysis* (Inuvialuit Communications Society 1992) that curated my news dispatches for the Inuvialuit newspaper. *Tusaayaksat*.

31 Details from the day of the hunt were originally reported by the author in the untitled front page commentary of Tusaayaksat, 20 September 1991, 1 that begins, 'History was made at Shingle Point'; and in 'A magical moment at Shingle Point' and 'The 1991 bowhead hunt: A chronology,' in Barry Zellen (ed.), The inuvialuit bowhead harvest of 1991: A pictorial history and analysis (Inuvialuit Communications Society 1992), 35.

32 As described by Inuvialuit Communications Society (ICS) television producers Debbie Gordon-Ruben and Stan Ruben in 'Tamapta's View of the Hunt,' *Tusaayaksat*, 20 September 1991, 2; and in Debbie Gordon-Ruben and Stan Ruben, 'A view to the kill', in Barry Zellen, *Bowhead Harvest*, 45.

33 The details presented in this section on the historic 1991 bowhead hunt were originally reported by the author in: Barry Zellen, 'Bowhead Harvest', 5–9.

34 The requirements of the Minister of Fisheries and Oceans are spelled out in Section 14 of the *IFA*. According to Sec. 14(65), 'Recommendations of the Fisheries Joint Management Committee pursuant to paragraph (64)(i) shall be forwarded to the Minister of Fisheries and Oceans, who shall implement, vary or reject them.' Sec. 14(66) explains that 'where the Minister of Fisheries and Oceans varies or rejects a recommendation of the Fisheries Joint Management Committee he shall provide the Committee with written reasons for his decision within thirty (30) days after the recommendation is made.' Sec. 14(67) stipulates that 'on receiving the decision of the Minister of Fisheries and Oceans to vary or reject a recommendation, the Fisheries Joint Management Committee shall consider the decision and within thirty (30) days submit a further recommendation to that Minister.' According to Sec. 14(68), 'on receiving the further recommendation of the Fisheries Joint Management Committee, the Minister of Fisheries and Oceans shall implement, vary or reject it.' Sec. 14(69) requires that 'where the Minister of Fisheries and Oceans varies or rejects the further recommendation of the Committee, he shall provide the Committee with written reasons for his decision within thirty (30) days after the recommendation is made.' *IFA*, Section 14.

35 A land dispute causing two fatalities.

36 Information in this section was originally reported by the author in: Barry Zellen, 'Congratulations bowhead hunters', *Tusaayaksat*, 7 October 1991, 6–7; and Barry Zellen, 'Celebrating the rebirth of a proud tradition', in Barry Zellen (ed.), *The inuvialuit bowhead harvest of 1991: A pictorial history and analysis* (Inuvialuit Communications Society 1992), 50.

37 Quoted on the inside back cover of Barry Zellen (ed.), *The inuvialuit bowhead harvest of 1991: A pictorial history and analysis* (Inuvialuit Communications Society 1992).

38 Jay Alabaster, "Kitchen car' whale cuisine: New dishes for a new generation', *Japan Forward* (Tokyo, 29 August 2022), <https://featured.japan-forward.com/whalingtoday/2022/08/29/kitchen-car-whale-cuisine-new-dishes-for-a-new-generation> accessed 23 April 2023.

39 Associated Press, 'Japanese company launches whale meat vending Hiroshi Terashima, 'Director's message' machines: Major supermarkets have largely stayed away from whale meat to avoid protests by anti-whaling groups', Food Manufacturing (Madison, 30 January 2023) <www.*foodmanufacturing*.com/facility/news/22684143/japanese-company-launches-whale-meat-vending-machines> accessed 23 April 2023.

40 Justin McCurry, 'Whale meat on the menu as Japanese suppliers try to tempt tourists: with the domestic market in long-term decline, whalers and restaurants are working with the Japan travel bureau in a bid to win over skeptical tourists', *The Guardian* (London, 23 March 2023) <www.theguardian.com/environment/2023/mar/23/whale-meat-on-the-menu-as-japanese-suppliers-try-to-tempt-tourists> accessed 23 April 2023.

41 Niall Alexander Rand, 'Reforming the International Whaling Commission: Indigenous peoples, the Canadian problem and the road ahead', (2017), 19 *ICLR* 2–3, 324.

42 Ibid.

43 Joanna Kerr, 'Greenpeace apology to inuit for impacts of seal campaign,' *Greenpeace. org,* 24 June 2014, <https://greenpeace.org/canada/en/story/5473/greenpeace-apology-to-inuit-for-impacts-of-seal-campaign> accessed 23 April 2023.

44 *Japan Forward,* 'INTERVIEW: Sustainable resource management is still the key goal for Japanese whaling – Tsutomu Tamura', Japan Forward, 30 October 2021, https://japan-forward.com/interview-sustainable-resource-management-is-still-the-goal-for-japanese-whaling-%e3%83%bc-tsutomu-tamura/.

45 Dennis Normile, 'Why Japan's exit from international whaling treaty may actually benefit whales: Commercial whaling will replace a controversial research program, but the market for whale meat is declining', (2019] Science, <https://www.science.org/content/article/why-japan-s-exit-international-whaling-treaty-may-actually-benefit-whales> accessed 23 April 2023.

46 Hiroshi Terashima, 'Director's message', *Ship and Ocean Newsletter*, Selected Papers No. 4, Institute for Ocean Policy, Ship and Ocean Foundation (SOF), an English-language version of papers from the Japanese Newsletter edition published from No. 41 (20 April 2002) to No. 50 (5 September 2002), 2.

47 Joji Morishita, 'Resumption of whaling and the principle of sustainable use', *Ship and Ocean Newsletter*, Selected Papers No. 4, Institute for Ocean Policy, Ship and Ocean Foundation (SOF), an English-language version of papers from the Japanese Newsletter edition published from No. 41 (20 April 2002) to No. 50 (5 September 2002), 10.

48 Ibid., 10.

49 Ibid., 11.

50 Kishigami, 49 (n 17).

51 Freeman, Wein and Keith (n 22).

52 Peter Varga, 'Nunavut whalers rejoice over successful bowhead whale hunt: Repulse Bay hunters land huge bowhead whale Aug. 31,' *Nunatsiaq News* (Iqaluit, 4 September 2013) <https://nunatsiaq.com/stories/article/65674nunavut_whalers_rejoice_over_successful_bowhead_whale_hunt/> accessed 25 April 2023.

53 David Venn, 'A successful hunt: How Baker Lake hunters got their bowhead; Maktaaq from the whale will be shared with the region's communities,' *Nunatsiaq News* (Iqaluit, 19 August 2021) <https://nunatsiaq.com/stories/article/a-successful-hunt-how-baker-lake-hunters-got-their-bowhead/> accessed 25 April 2023.

54 See map in Sellheim's and Morishita's chapter in this volume.

55 Justin McCurry, 'Japan's whaling town struggles to keep 400 years of tradition alive; The resumption of killing whales for profit for the first time in over 30 years is offering little cause for celebration', *The Guardian* (London, 26 December 2021) <https://www.theguardian.com/world/2021/dec/26/japans-whaling-town-struggles-to-keep-400-years-of-tradition-alive> accessed 25 April 2023.

56 Ibid.

57 Ibid.

58 Ibid.

59 Ibid.

9 Whales as 'sacred' and 'profane' in IWC Member State Cultures

Nikolas Sellheim

Introduction

The divide over what the International Whaling Commission (IWC) and with it the ICRW are to achieve was best exemplified at IWC67 in Florianópolis, Brazil, in September 2018. Here, two proposals were presented to determine the future of the IWC. On the one hand, Japan offered its 'Way Forward Proposal' which saw the conservation of whales along with a limited quota for a commercial whale hunt. On the other hand, Brazil and others offered the Florianópolis Declaration which saw the IWC move towards a more preservationist organisation and thus merely enabling Aboriginal Subsistence Whaling (ASW) and strictly no commercial whale hunts. The discussions over these competing proposals were heated and both 'sustainable use' countries and 'conservation countries' did not hold back in their accusations of the respective 'other side'. In the end, the Japanese proposal was voted down with 40 against and 27 in favour while the Brazilian proposal was accepted with the same inverted results.[1]

Forerunners in the debates were Japan and Brazil, both vocally accusing each other of dishonesty and failure to grasp the purpose of the IWC even though in other fields of governance the two countries maintain very good relations. Why, therefore, is the IWC in such a position that countries are unable to agree on whether or not to hunt whales? In this chapter, I explore the problem of cultural conflict as a reason for countries not being able to agree on a specific issue, such as whaling. The underlying hypothesis is that if countries, and thereby the IWC, were to approach whaling and anti-whaling through a more cultural lens, an entirely new way of communication might be possible. I therefore use Japan's pro-whaling attitude as well as predominantly Western countries' anti-whaling attitudes as an example and contend that either 'side' is neither right nor wrong, but that there is a fundamental lack of understanding – willingly or unwillingly – for the cultural dimensions of whales and whaling in those countries holding opposing views. Ultimately, therefore, this chapter makes the call for the establishment of a Committee on Cultures and Whales (CCW) within the IWC as a means to overcome, or at least discuss, deeply rooted cultural differences pertaining to whales and whaling.

DOI: 10.4324/9781003250814-11

Definitions and Methodology

While in this chapter, I frequently and regularly refer to the term 'culture', I am not implying that I have found a definition of the term. Yet, without having to delve into the literature on what culture is, could be or even should be,[2] I am very generally adopting Robert Axelrod's as well as Birukou's, Blanzieri's, Giorgini's and Giunchiglia's approach by assuming that the content of a culture is 'a set of traits, which can refer to behavior, knowledge facts, ideas, beliefs, norms, etc.'.[3] In this sense, in this chapter, I am not referring to *the* culture that considers whales and whaling in a certain manner, but rather whales and whaling as one specific cultural trait within a society.

Methodologically, this chapter is based on intense literature reviews on whales and whaling in different countries as well as the cultural perception of the environment and animal welfare. Moreover, the chapter rests on information collected through an online survey that was disseminated by the snowball technique through social media (Facebook, LinkedIn, Twitter) and the author's personal website (www.sellheimenvironmental.org). Lastly, the documentation of the meetings of the IWC of the last 50 years (since 1972) was coded using Atlas.ti. A coding scheme was constructed around interventions of countries and non-governmental organisations that have brought the cultural dimensions of whales and whaling – certain idea(l)s that accompany the taxon of cetaceans and their utilisation – to the fore.

Theoretical Background

This chapter deals with whales and whaling in light of Japan's withdrawal from the ICRW. The assumption that provides the theoretical underpinning of this chapter is that neither science nor conservation issues have in the end caused the breach between Japan and the majority of IWC members, but rather conflicts of cultural values pertaining to whales and whaling. This means that neither the IWC's Scientific Committee (SC) nor the plenary itself could have prevented Japan's withdrawal since both 'sides' (meaning states wishing to re-introduce a quota for commercial whaling and states favouring the preservation narrative) have consistently argued along lines of science.[4] While the decisions of the IWC concerning whaling are to be based on science, Heazle has demonstrated that while science is supposed to be 'neutral', it is nevertheless dependent on how scientific data is politically interpreted and used. Moreover, it very much depends on how stringently the precautionary principle is applied and on whose shoulders the burden of proof rests.

In the case of the IWC, the precautionary principle is indeed applied in a very stringent manner, meaning that the current interpretation of the moratorium is rooted in the assumption that given the unknown population status of some whale species and their sub-populations, it is better not to hunt them at all.[5] Japan, on the other hand, has time and again tried to demonstrate through its different scientific whaling programmes in the North Pacific and the Antarctic that certain populations do in fact withstand limited commercial whaling.[6] This also goes hand in hand with the so-called Revised Management Procedure (RMP), which was developed by the SC and adopted by the Commission in 1994.[7] In essence, the RMP allows for the establishment of catch limits by taking scientific uncertainty and risk into account. It has, however, never been put into action due to the failure of the Commission to adopt the Revised Management Scheme (RMS) and the accompanying limited lifting of the 'moratorium' on commercial whaling.[8]

The failure of the IWC to find consensus on whether the moratorium should be lifted or not, I argue, is consequently not based on scientific rigour or concerns over the population status of whales, but rather on deeply rooted differences over the way whales and whaling are culturally perceived. While for countries like Japan, whales are but one marine resource that can be exploited, for many Western countries whales have become enigmatic and to which certain narratives are attached: they are intelligent, social, can sing, can speak, are endangered and share many similarities with humans. In the recently released movie *Avatar: The Way of Water* by James Cameron, whale-like creatures, the Tulkun, are depicted exactly in this manner.[9] Without differentiation, 'the' whale is created – a 'species' that Norwegian anthropologist Arne Kalland has termed 'the superwhale'.[10]

The meetings of the IWC are consequently not spared from these different narratives and national governments follow different agendas in terms of their way of arguing for or against whaling. Consequently, what we face in the IWC is not necessarily a matter of scientific findings, but rather a failure of intercultural communication or intercultural dialogue. In other words, the IWC has failed to reconcile the different cultural values that are attached to whales and whaling.

This leads to the assumption that in international organisations, such as the IWC, states cannot be considered purely rational actors. Here, 'rational' is to be perceived not in the sense of staying in an agreement when the benefits outweigh the costs (or leaving it when this is not the case anymore),[11] but rather in the sense that the agreement's external effects do not directly affect certain states or not ever having directly affected them: in the case of the IWC, there are many states – if not the majority – that have never had a whaling history and have joined the IWC just in order to further a specific cause. Sweden and Finland, for instance, have joined the IWC exclusively with the intent to support the moratorium but without having a whaling history.[12] In this sense, the IWC has far exceeded its optimal membership since states without a whaling history are now able to make decisions on equal footing as those that have one. This, as a consequence, hinders the efficiency of the organisation by exceeding its externalities and instead leads to normative clashes on whether or not whales can or should be used for human consumption.[13] Exemplified by the comparatively small number of Parties participating in the meetings of the SC *vis-à-vis* the number participating in the IWC plenary meetings, this allows for the conclusion that the majority of states are more interested in the normative aspects than scientific aspects of whaling.

Inevitably, this situation leads to a numbers game: it needs ¾ of the IWC members who are present at the meeting and cast a positive or negative vote in order to amend the Schedule, which then could lead to a lifting of the moratorium. This forces states that have voted against a certain measure, such as the moratorium, to comply with it anyway since they are required by the respective treaty to comply with majority decisions. Robert Friedheim contends that in the IWC this consequently corresponds more to a legislative body than to equal partnership under international law.[14] As a consequence, in order to make one 'side' 'win' in this situation, strong states are left with several options: first, to win states over by argument, for instance based on science; second, to encourage like-minded states to join the IWC so that the numbers are finally reached; or, third, through coercion or 'vote-buying'.[15]

Needless to say, these options are all rather confrontational and not in the spirit of cooperation and consensus as in other international organisations. In each of these options, what is completely left out is intercultural dialogue or cultural communication in order to find reconciliation for adversarial stances on whales and whaling and to find a mutual understanding of the role whales play in a respective society. In essence, therefore, this is a matter of understanding 'sacred values', in order to further the pluralistic idea(l)s of international law and decision-making. Because, in the end, there is an inherent clash of values concerning whales: while pro-whaling states argue along lines of culture, necessity (or needs, as in the case of aboriginal subsistence whaling) or even benefit that must be gauged against conservation statuses, anti-whaling states locate whaling within a precautionary and therefore moral dimension. For them, hunting whales is as such not possible since whales are sacred in so far as they are not to be considered as a food source. The rejected resolution on food security at IWC67 is one example in this regard.[16]

We consequently enter the domain of the 'sacred' and the 'profane', which Emile Durkheim has touched upon in *The Elementary Forms of Religious Life*.[17] For some, whales are somewhat untouchable, corresponding to the overarching will of society and subordinating humans to some degree ('sacred'). For others, whales serve as a food source or even as a means for financial gain, for fulfilling 'profane' human needs. But by aiming to re-open commercial whaling, the profane touches the sacred, which, according to Durkheim, in this dichotomy of human perception, cannot happen without impunity.[18] The scales by which these are measured are inherently different and cannot be linked easily: there are things that simply go beyond material or other trade-offs. For example, the parental instinct to protect your children, freedom movements that ignite a revolution, such as the so-called Arab Spring,[19] or the cause of a suicide bomber are measured by different scales – scales beyond the tangible – than economic benefit, food security or other cost-benefit justifications.

For the purposes of this chapter, therefore, I consider whales through the lens of the 'sacred' and the 'profane'. While, as a point of departure, I consider that this divide exists in the IWC and in the different cultures represented in the IWC, it becomes ever more necessary to establish a dialogue between the different 'camps' in order to understand one another, to overcome the divide and to work on a mode of communication that goes beyond the tangible and beyond science.

Discourses on the Environment and Concerns for Animals

It is noteworthy that the dichotomy between the 'sacred' and the 'profane' in a whaling context is more than simplified, however serving our purpose to justify the establishment of a Committee on Cultures and Whales. Studies on cultural and perceptional dimensions of animal welfare or environmental protection, however, have demonstrated that even within a specific country, there is no unison view on these matters, having generated different discourses that take hold. Kellert, for instance, considered discourses on animal concerns through ten different typologies:

– *Naturalistic*, with a major affection for and primary interest in wildlife
– *Ecologistic*, with a systemic view on and primary concern for the natural environment as well as the different interactions
– *Humanistic*, with a major affection for and primary interest in individual animals
– *Moralistic*, concerned with a moral approach (right or wrong) to animal treatment, but with a strong opposition to cruelty to animals
– *Scientistic*, with a primary interest in the physical and biological parameters relating to animals
– *Aesthetic*, holding primary concern for the artistic and symbolic characteristics of animals
– *Utilitarian*, with a primary concern for the practical and material values of animals and the ecosystem(s)
– *Dominionistic*, with a primary interest in the mastery and control of animals, particularly as regards sporting events
– *Negativistic*, primarily interested in an active avoidance of animals because of dislike or fear; and
– *Neutralistic*, primarily depicting the avoidance of animals due to lack of interest or indifference.[20]

When considering the discourses within the IWC, the 'sacred' view on whales would comprise the naturalistic, ecologist, humanistic, moralistic, scientistic and aesthetic elements of Kellert's typologies. On the other hand, the 'profane' view on whales would comprise scientistic and utilitarian perceptions. To some degree, a dominionistic perspective could be referred to, especially when observed interventions at IWC meetings stress the natural law view of a dominion of people over nature. This, however, is not the norm.

In a later study, Kellert was able to specify his findings through a case study referring to Germany, the United States and Japan. In this study, Kellert found that in Germany and in the United States, it was especially 18–35-year olds and those with a better education that showed greater concern for wildlife and greater interest in them. Generally, however, respondents in these countries were more concerned and interested in pets instead of wildlife.[21] Linda Kalof, building on Kellert and other older studies,[22] was able to discern six distinctive discourses on concerns for animals.

– Animals as loved ones, to be shielded from extinction and holding basic rights
– Animals as serving humans for animal products, recreation, but shielded from overexploitation
– An anthropocentric view towards animals with a primary focus on human well-being, even at the cost of animals

- A disconnect from animals, but a great concern for wildlife
- A focus on pets, but with a degree of disconnect from wild animals and animals that are considered 'pests' by humans
- An ecocentric view that considers the protection of wildlife more important that human development[23]

As the above already indicates, there is no one-off attitude towards conservation of species in either of the examined groups. Political attitudes, age as well as gender play a role in shaping the moral view towards environmental conservation, also dependent on the species (both singular and plural) that are to be protected. In addition, emotional responses, scientific names and a species' role in the ecosystem contribute to attitudes towards conservation in different age groups, as Kelemen et al. have found,[24] while the degree and modes of participation,[25] the 'cleanliness' of public spaces[26] or vague narratives about a species[27] contribute to shaping discourses on conservation and/or utilisation. Against this backdrop, it is difficult to identify a geographically situated attitude towards conservation. The European Union's argument for a morality-based ban on trade in seal products before the World Trade Organization is consequently difficult to uphold, as demonstrated elsewhere,[28] making it furthermore difficult to argue within a body such as the IWC to speak for a country's people when arguing for or against the lifting of the whaling moratorium.

This notwithstanding, certain *political* discourses exist that appear to make use of cultural elements. For instance, in Japan, politically whaling is a matter of cultural identity and therefore finds support from a wide array of Japanese political parties, represented in the National Diet of Japan.[29] This does not mean, however, that public attitudes towards whaling in Japan necessarily support this view, as will be shown below. Be that as it may, there still appears to be a rift between representatives within the IWC concerning the use and conservation of whales. For the sake of argument, these are denoted as 'profane' and 'sacred', following Durkheim.

The 'Profane' Whale

When the IWC took up its work in 1948, the *zeitgeist* concerning environmental protection, species conservation and international cooperation on these very issues was fundamentally different to what it is today. As I have shown elsewhere, it was precisely the exploitative nature of human environmental perception that fuelled international agreements on species conservation. In other words, species, and particularly marine mammals, were perceived through a lens that made them a subject of human benefit for which they are (or were) to be protected rather than being inherently valuable as sentient beings.[30] As a consequence, early day multilateral agreements were, as per Durkheim's dichotomy, located in the realm of the 'profane' from today's perspective. From the perspective of the drafters of the ICRW, this was not the case at all, however.

Because at that time, whales had not taken up the same societal position in public discourse as they have now and those seven states that agreed on the ICRW –

Australia, France, Norway, the Soviet Union, South Africa, the United Kingdom and the USA – are retrospectively considered a 'whaling club'.[31] The now much disputed objective of the ICRW, and thereby the IWC, was consequently rather clear during that time: 'the proper conservation of whale stocks and thus make possible the orderly development of the whaling industry'.[32] Whales should be protected so that the whaling industry could thrive.

It therefore does not come as a surprise that the records of the IWC do not indicate a significantly different approach among its member states to whales and whaling during the first years of the IWC's existence. While already in the 1950s, it was time and again stressed that the Commission's purpose is to protect whale stocks (e.g. in its third report, the IWC viewed "with some disappointment the continued failure of countries interested in whaling to co-operate in the maintenance of the principles underlying the 1946 Convention on which the future conservation of the whale stocks of the world so much depends"),[33] it was nevertheless the dominating blue whale unit (BWU) and the maximum exploitability of whales that steered its work. For example, in its fourth report, the Schedule amendment for the 1953 season concerning humpback whales shows that if within some given days the maximum number of 1250 humpback whales had not been reached, the whaling season could be extended by another few days until that maximum was reached.[34] Maximum exploitability consequently drove the management of whales at that time.

Generally, it can be said that throughout the 1950s whales were considered a resource that was to be exploited, albeit taking conservation statuses into account. This was usually measured by the BWUs *vis-à-vis* the different types of whales that were landed at the end of each season. The Commission was concerned that certain whale species, such as the blue whale in the North Atlantic, were subject to overly great harvests, prompting its members to either establish sanctuary areas or prohibitions on whaling. All this against the backdrop, however, that whales should continuously serve as an exploitable resource. This is further underlined by the first application of 'special permit whaling' by several nations. As per Article VIII of the ICRW, states can issue special permits for whaling for scientific purposes. While in recent years, i.e. in the 1990s and 2000s, this was a matter of massive controversy, in the 1950s, special permits were issued by Canada, the Soviet Union, the United Kingdom and the Netherlands without any resistance from other IWC members.

While this may be so, the eighth report of the Commission, published in 1957, already hints towards an emerging rift between Japan and other members of the Commission. Here, the issue concerned the prohibition of the taking of blue whales in the North Pacific, which was not upheld, also by Japan. Japan argued that the prohibition had lasted for more than 15 years and that it appeared as if the stock had increased. This notwithstanding, Japan limited its annual catch to 70 individuals and was in the process of conducting scientific studies concerning the status of blue whale stocks. Only if this investigation was to support international limitations on blue whale takes was Japan able to accept these limitations.[35] Similarly, Iceland – and with it also Denmark – refused to withdraw their objections to the five-year protection of blue whales in the North Atlantic, of which '[t]he Commission deeply regretted to learn'.[36] While the report does not provide any reasons for this refusal, it becomes clear that Japan, Iceland and Denmark act against the prevailing view of the Commission based on their own take on certain whale (stocks).

It was not until 1958 that animal welfare aspects were considered. Contrary to the Canadian commercial seal hunt where already in the 19th century the wastefulness and welfare of the seals were criticised,[37] at least in the IWC humane killing methods were put before the Commission and became an agenda item for the 1959 meeting. Since then, the issue had become part and parcel of the meetings different agendas. While states like Japan, Norway or Iceland quite substantially pushed for higher BWUs and the maximum exploitation of whale stocks, it is especially Norway which was in the process of developing a new harpoon to make the killing of whales as painless as possible. In order to do so, it closely cooperated with other members of the Commission. This being said, since the allocation of the BWU in Antarctic whaling could not be found a consensus for, the Netherlands and Norway announced their withdrawal from the Convention as of 1 July 1959. While also Japan announced that it would withdraw, this announcement was revoked in the end.[38]

With the different interests of the whaling nations coming together in the IWC, it was at this time that these interests started to collide. While all IWC member states at that time did not question the purpose of the IWC as such – after all, the so-called environmental revolution had not taken hold of the world's societies – the organisation was nevertheless already then marked by disagreements over how to proceed. On the one hand, the BWU was considered the guiding unit for whaling operations. On the other hand, it was not considered flawless by many. This, in the end, let Norway and the Netherlands to (temporarily) leave the Commission.

The following years were marked by scientific guidance being the benchmark of any whaling operations. While the BWU still steered the way the SC provided its advice, it corresponded to the scientific status quo of that time, which meant that advice was provided based on the best scientific knowledge. Whether or not this was reasonable and in the interest of whales is a different matter.

This notwithstanding, in 1962, Rachel Carson published her seminal book *Silent Spring*,[39] which led to an increased awareness of the potentially negative impacts of human activity on the natural environment based on the use of pesticides, and especially Dichlorodiphenyltrichloroethane (DDT), in the United States. While it is probably somewhat oversimplified to accredit Carson with the launch of the environmental movement,[40] it is nevertheless an important capture of the prevailing increasing concern for environmental issues that had taken hold of Western societies – not least because of the spread of television and the repeating presentations of environmental problems.[41] Moreover, the 1960s were marked by the time when the 'Space Race' prompted the first images of Earth. One of the most iconic images in this regard was taken by Apollo 8 in 1969, entitled 'Earthrise', which shows the surface of the Moon over which Earth rises. This image shows the fragility of Earth against the backdrop of the emptiness of space. In 1969, the first humans set foot on the Moon and in 1972, another iconic image was taken: 'The Blue Marble', which is the first image of Earth taken from space, which shows the entire planet. The impacts of these images on humankind's perception on the environment cannot be underestimated, paired with the narratives of the early astronauts, who frequently expressed their own fragility and the fragility of 'spaceship Earth' that they saw when they first went to space.[42]

In other words, throughout the 1960s, a shift particularly in Western societies took place, resulting in the first World Earth Day in 1970, proposed by the American Democratic Senator Gaylord Nelson – a proposal that more than 20 million Americans would follow, taking to the streets and demonstrating for a cleaner environment and more environmental protection.[43] At that time, at the 23rd meeting of the IWC in 1971, the Schedule was amended to take into account the diverse interests of the whaling industry, which, for instance, had to maintain two types of factory ships: one for the Antarctic and one for the North Pacific. While catch limits had been established for baleen whales also for the North Pacific, subject to these catch limits also Antarctic factory ships could be used in that area. Moreover, at a proposal from Norway, minke whales were excluded from regulatory measures in the Antarctic.[44] It is important to mention that at that point, out of eight species, four were fully protected: blue, gray, right and humpback whales. For the other species, sei, fin, sperm and minke whale, other measures, such as closed areas, closed seasons and size limits, were put in place in order to conserve their stocks.

However, at the same meeting, divergences between the interests of the whaling industry and the SC became apparent: in order to fulfil the needs of the whaling industry, Japan and the USSR argued for a higher maximum sustainable yield (MSY) for Antarctic fin whales than the SC originally proposed. Both Japan and the Soviet Union considered the population estimates as too low to be realistic and they were the only countries in the Commission arguing that the population of Antarctic fin whales had increased. They relied on their own science, rather than the findings of the SC.[45]

It was also IWC23 that marked a significant shift 'towards improving the effectiveness of the Commission's measures to ensure proper conservation of the whale stocks'.[46] Because it was at that meeting that the BWU was abandoned and replaced by a species-based allocation of quotas starting with the 1972/73 whaling season. The replacement of the BWU was first proposed by the United States already in 1969 and finally recommended by the SC.

In 1972, the first *United Nations Conference on the Human Environment* took place in Stockholm. While setting important standards for environmental diplomacy by establishing the United Nations Environment Programme, the 109 recommendations of the conference also included a recommendation to strengthen the IWC, to strengthen research efforts on cetaceans and to impose a ten-year moratorium on commercial whaling.[47] The United States, seconded by the United Kingdom, tabled a proposal in this regard at IWC24, wishing to amend the allocation of quotas to zero in the Schedule for the next ten years. They argued that a ban on commercial catches would appear sensible in light of the insufficient knowledge on whales. Others, however, argued that due to the important scientific findings that derive from caught whales, research on whales would suffer from such a ban. Also the SC noted that a 'blanket moratorium is in the same category as a blue whale unit quota, in that they are both attempts to regulate several stocks as one group whereas prudent management requires regulation of the stocks individually'.[48] In the end, the proposal was rejected with four in favour, six against and four abstaining.[49]

The 1970s were marked by ongoing discussions on a moratorium on commercial whaling. During that time, states such as Finland and Sweden joined the Commission, neither of which having a history of whaling. Their explicit goal was to put a moratorium in place. At IWC34 in Brighton, England, all in all five proposals, tabled by the Seychelles, the UK, the USA, France and Australia, were brought forth that sought to end commercial whaling. In other words, the discourse on a 'sacred' whale started to take hold. In the end, the Seychelles proposal that sought to amend paragraph 10 of the Schedule, putting zero-catch-quotas in place, was voted upon: 25 votes in favour, with 7 against and 5 abstentions.[50] Those states that voted against the moratorium were Japan, Norway, Iceland, South Korea, Peru, Brazil and the Soviet Union.[51]

From the very beginning, several states lodged objections to this amendment: Norway, Peru, Japan and the Soviet Union. This meant that despite the moratorium, they were still able to hunt whales commercially. Iceland did not object to the moratorium, but instead used the 'scientific exception' to obtain whale meat. In the years to come, Peru and Japan withdrew their objections, meaning that from then on they were also bound to the moratorium. In Japan's case, however, this did not occur because of a sudden change in perception, but rather because of political pressure, particularly exerted by the United States and a looming certification under the 1979 Packwood Magnuson Amendment to the Magnuson-Stevens Fishery Conservation and Management Act. Under this amendment, the US government can reduce fisheries operations of a foreign state in the United States waters by 50% in case this state violates the effectiveness of the ICRW. In light of this threat, Japan was forced to withdraw its objection. As a consequence, it started to focus on scientific whaling operations.[52]

The adoption of the moratorium probably marks the most important turning point at the IWC. Without having to go into detail, the split between states opposing whaling and those states in favour has caused a rift in the Commission that cannot be overcome and that, in the end, has prompted Japan to leave it altogether. The scales of difference between the 'profane' and the 'sacred' appear irreconcilable. While the Russian Federation (as the successor of the Soviet Union) and Norway have not withdrawn their objections to this day, they are not bound to the moratorium – a right that Norway makes use of in the North Atlantic. Iceland solved its opposition to the moratorium by leaving the IWC in 1992 when it launched, along with Norway, the Faroe Islands and Greenland, the North Atlantic Marine Mammal Commission (NAMMCO) – an organisation that supports the principles of conservation and sustainable use. In the North Atlantic, therefore, despite the developments elsewhere in the world, whales have remained species that are eligible for hunting and that serve as a source of food. 'The whale' has thus maintained its 'profanity' since there is no reason, culturally and politically, for NAMMCO countries to change their positions.

Taking a closer look at Japan, it becomes quite obvious that whaling is a rather integral, 'normal' part of Japanese society. Apart from exquisite whale restaurants in Tokyo, Osaka or Taiji, whale meat can be found in supermarkets and in small *izakaya* restaurants – albeit often merely upon request. There have been different surveys on Japanese (dis)approval of whaling. A survey commissioned by the International Fund for Animal Welfare found in 2012 that almost 27% supported lethal scientific whaling, 18% opposed it and the rest was undecided.[53] Two years later, the *Asahi Shimbun*, one of Japan's leading newspapers, conducted a survey that found that 60% of the Japanese supported whaling even though the minority had ever tasted whale meat.[54] No recent surveys or polls exist. This notwithstanding, it is clear that Japan has had a long-standing history of whaling. While the commercial whale hunt has started after the Second World War as a means to counter the massive food shortage, coastal whaling has been an ongoing activity for centuries. The deep cultural roots of whales and whaling have been documented on numerous occasions.[55] They can, for instance, also be found in numerous temples and shrines all over the country, along with ceremonies that mourn the whales after they had been killed.[56]

The survey that was carried out for the purposes of this chapter generated 113 responses from Canada, Germany, the United States, France, Switzerland, Norway, Finland, Australia, Austria, Brazil, Sweden, Iceland, Greenland, the United Kingdom and South Africa. It shows that the general public appears to be somewhat leaned towards an opposition of whaling whereas 22.5% of the respondents are strongly opposed to whaling, 28.96% oppose whaling, 11.46% do not have a view, 22.83% endorse whaling and 11.24% strongly endorse whaling (see Figure 9.1).

The majority of respondents consider whales as a species that is not to be killed (40.8%), 33.85% see whales as a resource, 20.72% consider whales as neither and 4.65% do not have an answer (Figure 9.2).

64.58% of the respondents furthermore note that in their respective home countries, whales are considered as intelligent, social and endangered, 21.04% consider the discourse to present whales as a source of food and other resources where as 14.38% see the discourse as presenting whales as neither in their respective home country. The majority of respondents do not watch programmes on Animal Planet, Discovery Channel or similar outlets (56.66%) whereas 43.34% do, indicating that the information provided in the respective programmes may also help shaping an opinion. In an open question, respondents showed that the primary source of information concerning whales and whaling stems from the internet and news outlets. Only a very minor number referred to scientific studies as their source of information whereas even fewer made reference to the IWC as a source of information. Merely 16.27% think that whaling is illegal under international law, 58.51% think that it is legal while 25.22% do not know (Figure 9.3).

While the survey did not follow a special methodology nor is it representative, it nevertheless gives at least an indication for the overall attitudes of those people interested in the issue of whales and whaling. From the survey, it therefore appears that the 'sacred whale' is not necessarily the prevailing narrative when it comes to whales and whaling.

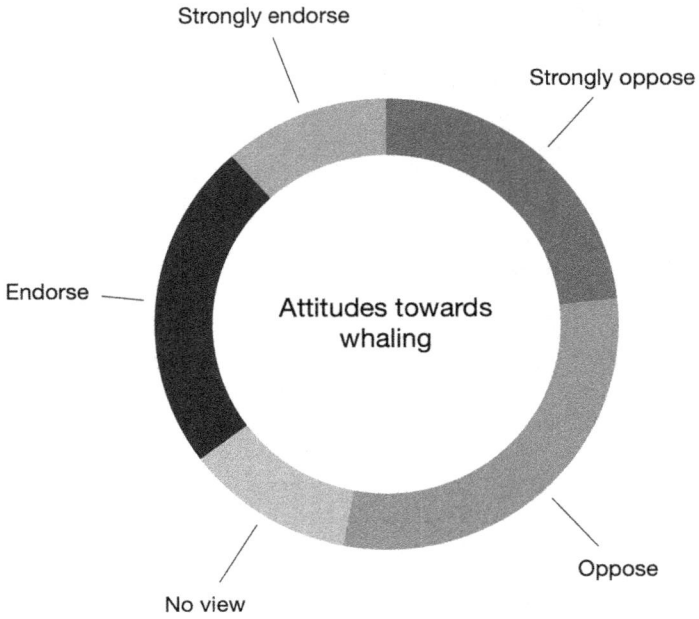

Figure 9.1 Attitudes towards whaling.

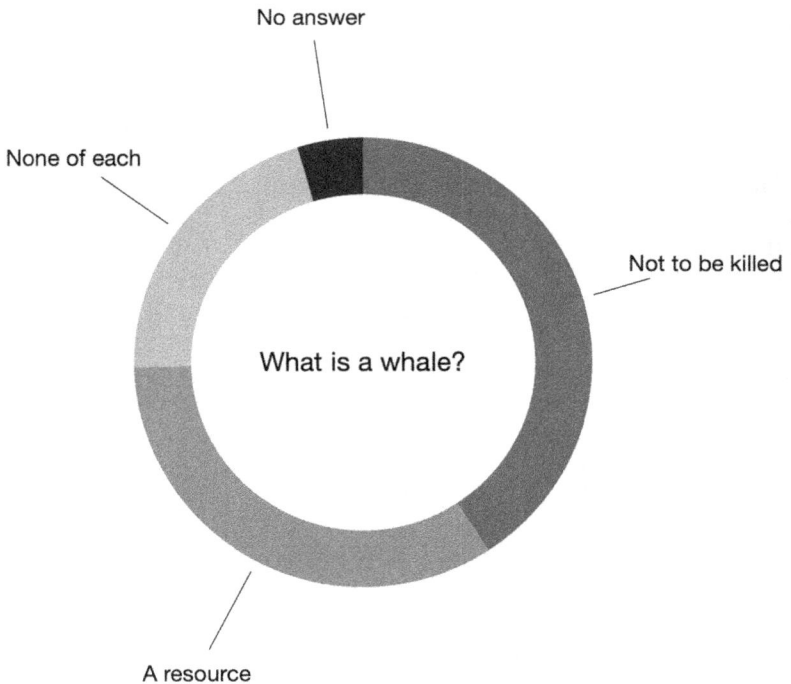

Figure 9.2 What is a whale?

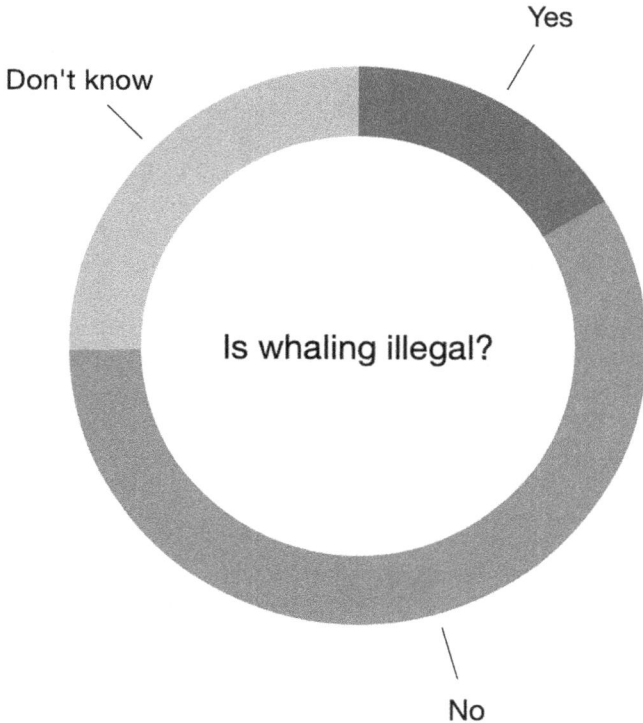

Figure 9.3 Is whaling illegal?

The 'Sacred' Whale

When the IWC was established, the idea of a 'sacred' whale was non-existent. As we saw in the preceding section, only over the course of the post-war period, environmental concerns and an entirely changing paradigm started to emerge, which also translated into the whale becoming an iconic species rather than a resource.[57] One of the early cultural reflections of this changing paradigm was Judy Collins' 1970 album *Whales and Nightingales* and her interpretation of the 1850s whaling song 'Farewell to Tarwathie' – a song which carried a melancholy that seems to condemn the very act of whaling.

The changing paradigm could also be seen in the massive resistance against commercial sealing in Canada. The strict positioning of celebrities, e.g. Brigitte Bardot, against the killing of seals, the images that started to fill the ever-growing numbers of TV screens in Western societies and the strong political messages that were carried to the wider public by NGOs symbolise a changing *zeitgeist* during that time.[58]

The US Marine Mammal Protection Act of 1972, which outlawed the commercial take of all marine mammals in US waters as well as the decision of the then-European Community to ban the import of whale products were a signal that

showed which way the large economies of the world would go.[59] This is not surprising. After all, in 1970, the 'superwhale' saw its discursive formation, especially when the bioacoustician Roger Payne recorded and released the *Songs of the
Humpback Whale* in 1970, heavily influencing the 1972 Stockholm Conference
and the 'Save the Whale' movement. Just a few years later, in 1979, *National
Geographic* produced an issue that contained a record with excerpts from the 1970
release. This was sent to 45 million subscribers. Moreover, the Voyager probes
that were launched in 1977, which carry the so-called Golden Record that contains
sounds and languages from Earth, contain excerpts of the *Songs of the Humpback Whale.*[60] In Germany, in 1991, issue 6 of the German version of the *Mickey
Mouse* magazine – as shown in Figure 9.4. – contained a flexi disc record, which
held 'songs of [humpback] whales' and contained a large print that read 'save the
whales!' ["Rettet die Wale!"]. The narrator of this 3+ minutes record notes that the
singing enables whales to talk to each other, but 'unfortunately we don't understand them yet' ["leider verstehen wir sie noch nicht"]. This notwithstanding, we
still know that it is a love song that they sing, because they meet (at a particular
island) to mate. The narration continues:

> When in earlier days the people went out with their boats to whale, they
> plugged their ears in order not to hear the singing. For who heard the whales
> signing out at sea, so it is said, can never again murder a whale.
>
> ["Wenn die Menschen früher mit ihren Booten zur Waljagd ausfuhren,
> verstopften sie sich die Ohren, um den Gesang nicht zu hören. Denn wer

Figure 9.4 Issue 6 of the 1991 German Mickey Mouse magazine, including flexi disc.

einmal die Wale draußen auf dem Meer singen hörte, so sagt man, der kann nie wieder einen Wal ermorden."].[61]

The anthropomorphisation of whales was therefore linked with a very prominent children's character – Mickey Mouse – and taken into the rooms of millions of children.

As a consequence, throughout the general public, the image of 'the' whale fulfilling certain character and ecological trades appeared. As mentioned above, 'the' whale was to become intelligent, friendly, musical, social and cultural. In addition, the whale was endangered and standing representative for the destructive impacts of humankind on the environment. A differentiation between species or stocks did not occur and, in fact, still does not seem to occur in public discourse.

Media-savvy organisations such as Greenpeace started to launch their first direct action campaigns against whaling fleets in the 1970s. For instance, in 1975, the Greenpeace vessel *Phyllis Cormack* confronted the Soviet whaling fleet in the North Pacific. What was new was that the activists put their own lives at risk in order to protect the whales from the whalers. This, quite obviously, caused significant media attention and put the plight of whales onto the public's agenda.[62]

Throughout the meetings of the IWC, this socio-cultural development cannot necessarily be traced. That said, the United States, also in light of its own national legislation, became a forerunner for the adoption of a whaling moratorium. Quite interestingly, the first reference to a moratorium can be found in the 23rd Report of the Commission,[63] where it was rejected due to its scientific infeasibility. Indeed, the SC, also in the years to come, defended the view that a moratorium could not be justified scientifically.

Despite the advice of the SC, the United States and the UK held a different view. At IWC24, the United States tabled a first proposal for a moratorium, seconded by the UK.[64] This proposal, however, was rejected by the Commission. As was shown above, especially the United States has been spearheading the environmental movement, which fuelled international debates and infused ethics on the human-environment relationship. It is therefore not surprising that even though the proposal was rejected, the discourse on the moratorium solidified within the Commission. The SC was therefore in a position that had to gauge political interests that became ever more prominent against the scientific reasoning that was to underlie decisions of the Commission.

At IWC25 in 1975 the United States, seconded by Mexico, tabled a proposal for a resolution that aimed for a ten-year moratorium – in many ways resembling the proposal tabled one year earlier. While the Technical Committee, a management committee of the IWC, approved the proposal, the SC reiterated its scientific infeasibility and also the Commission did not accept it as it stood. Australia – now a leading anti-whaling IWC member – and Denmark, on the other hand, tabled a proposal for a resolution that took into account the 'need to preserve and enhance whale stocks as a resource for future use and taking into consideration the interests of consumers of whale products and the whaling industry'.[65] The resolution differed in so far from the United States proposal in that it did not rule out commercial whaling, but rather aimed to distinguish each whale stocks into three distinct categories:

1 Initial management stocks: Permitted to commercially exploit following the advice of the SC, to a degree to bring it down to MSY level and then optimum level, but not below.
2 Sustained management stocks: Permitted to commercially exploit, following the advice of the SC.
3 Protection stocks: Fully protected stocks with no commercial whaling.

Even though the US Commissioner still opted for a ten-year moratorium, the resolution was accepted with the proposed additions.[66] This should have put the matter to rest. However, internal pressure within the IWC grew, especially when countries that explicitly opposed whaling started to join the Commission for the very purpose of pushing through a moratorium. Sweden and Antigua & Barbuda, for instance, joining the IWC in 1979 and 1982 respectively, had this goal (with Antigua & Barbuda now being a forerunner for the sustainable use of whales). The IWC, therefore, started to undergo a paradigmatic shift – in so far as it started to become an organisation that was no longer able, or willing, to manage whaling, but rather, the members of which aimed to protect whales from any lethal use. For many states, whales were no longer a 'profane' resource but instead had become an epitome for the sacredness of nature.

When the moratorium was finally put in place in 1982, the international response was positive and voices in the press presented it as a victory.[67] With the moratorium, the international discourse on whales and whaling changed even more. From a very personal perspective – growing up in Germany in the 1980s and 1990s – it was always clear that whaling was illegal since the IWC had made it illegal. As substantiated by the *Mickey Mouse* magazine, whales should simply not be killed. That the legal conditions for whaling were and are significantly more complex did not occur to me. After all, at that time, the IWC had a membership of 37 states while the UN had a registered membership of 158 in 1983.[68] Meaning, the moratorium on commercial whaling was merely binding for those states being a party to the ICRW and not having objected to it – as is still the case today.

When I drive down memory lane even further, 'the whale' had also become a species that was marked by the characteristics that Kalland[69] describes. In the Western, central European world, whaling was a no-go since it was unimaginable to consume whales. Instead, whales became poster motives on walls, cute toys that children could play with and icons for a better world, again best exemplified by the *Mickey Mouse* magazine. 'The whale's' majestic nature, its intelligence and close family ties triggered the imagination of millions. It is therefore not surprising that the whaling moratorium can be considered one of the most groundbreaking decisions an international environmental organisation has come to. The mere political nature of the decision, i.e. the elevation of 'the whale' onto a level that goes far beyond science, is probably best demonstrated by the response and resignation of the chair of the SC – Philip Hammond – to the rejection of the adoption of the RMP in 1993 even though the SC unanimously recommended the adoption. Hammond said that he was not able to be 'the organizer of and the spokesman for a Committee which is held in such disregard by the body to which it is responsible'.[70]

Up to the day of writing, the narrative of whales fulfilling Kalland's characteristics has prevailed. In 2008, the discourse evolved even further when the TV show *Whale Wars* found its way into the living rooms of the West. In this show, cameras accompanied the different campaigns by the direct action NGO Sea Shepherd in the Southern Ocean, facing Japanese whalers, which were themselves carrying out the scientific programmes as prescribed by the Institute of Cetacean Research in Tokyo. The show presented Sea Shepherd as David (standing representative for the whales), fighting evil Goliath (Japan) who is yearning for whale blood. Of course, this seemed to necessitate direct action by 'brave warriors' who risked their lives to counter Japanese aggression against whales.

The importance of *Whale Wars* as a cultural tool should not be underestimated in so far as it transports a certain message of a social movement as well as further pushing a profit-oriented ideology.[71] Looking more closely, however, the show boosted the TV station's Animal Planet viewership by 15%. On the other hand, even though it inherently circled around Sea Shepherd's crews rather than whales, it presented Japanese whalers as distant, two-dimensional beings and 'maintains the Japanese as the intractable other who destroys nature and lacks compassion'.[72] Any kind of justification of whaling, a Japanese view on the issue or simply any kind of attention that is given to the Japanese is fully lacking. Moreover, Japanese whaling vessels are presented as a destructive industrial force that drives its merciless teeth deep into the pristine wilderness of the Southern Ocean. As Robé asserts, this 'is an important first step in challenging the abstract rhetoric of industrialism, capitalism, and scientific rationality that the whalers employ to justify their slaughter'.[73] This narrative is pushed further in 2023 in *Avatar: The Way of Water*, which presents those that hunt the whale-like Tulkun as merciless, reckless and yearning for profit.[74]

Against this backdrop, it is not surprising that non-whaling states have solidified their anti-whaling stance even more. Even though Sea Shepherd has been banned from the meetings of the IWC since 1982 due to their aggressive actions against whaling vessels in Reykjavík harbour,[75] the majority-held opposition towards whaling within the IWC *vis-à-vis* those states still in support of it has led to the outside perception that whaling as such was no longer favoured by the organisation. Parties such as Australia, New Zealand or the member states of the European Union have become outspoken and media-savvy anti-whaling states – creating the impression that apart from Japan, Norway and Iceland, the IWC had become an organisation for the protection of whales and not for the management of whaling. The German Ministry of Food and Agriculture, for instance, writes on its dedicated website for 'whale conservation' that the IWC 'focuses on, inter alia, the conservation of certain whale species, whale sanctuaries, whaling quotas and whale research'.[76] In fact, when Iceland tabled its application for EU membership in April 2010, the German parliament agreed that one of the prerequisites for Iceland's accession to the EU would have to be the abandonment of commercial whaling.[77] The elevation of 'the whale' on to a level of the 'sacred' is consequently well established in Germany. But it has also become a narrative that is found time and again in different outlets up to the present day, substantiated by the majority of IWC members who vividly, actively and successfully oppose commercial whaling and thereby a lifting of the moratorium.

Reconciliation of Scales Irreconcilable?

Against this backdrop, it does not come as a surprise that within the IWC inherent cultural conflicts persist. On the one hand, Western countries such as EU member states, Australia and New Zealand along with Latin American countries (the so-called Buenos Aires Group) vehemently oppose any lethal utilisation of whales. This means that whales are removed from discourses of food security or cultural uses of whales. The only exception here is aboriginal peoples, for whom a special category of whaling exists in order for them to fulfil their cultural and nutritional needs – Aboriginal Subsistence Whaling (ASW). It is consequently not surprising that the Commission has thus far rejected any proposed resolution on food security, last tabled by Gambia, Guinea, Cambodia and Antigua & Barbuda at IWC68 in October 2022. This resolution aimed to establish at least the potential to use whales for food.[78]

The lenses through which whales and whaling are perceived in the Commission are scientific, on the one hand, and emotional, on the other hand. While science plays a major role in the deliberations of the Commission, as Heazle[79] has shown, science itself may be neutral, but the scientists conducting the science are not. The precautionary approach consequently plays a central role in the way cetacean science is politically applied in the Commission and how the attitudes of decision-makers can be swayed in a certain direction. This does not necessarily take place at the meetings of the Commission, but rather in the intersessional periods during which campaigns – either public or 'behind the scenes' – influence the position IWC member states will take during the meetings.

Japan's withdrawal from the Convention and therefore from the IWC is certainly a step in the wrong direction as this essentially demonstrates a failure of international diplomacy. Given the country's vehement push to obtain small quotas for commercial whaling up to its leaving, it is not unrealistic that other countries that face similar frustration (from their perspective) will follow suit in the future. Indeed, what is the point of remaining in an international body when one's own wishes are constantly overruled? Of course, from the perspective of those states opposing commercial whaling, the IWC functions well. After all, they have the majority of like-minded states behind them. Yet, it should also be in their interest to avoid states leaving in the future in order to prevent the potential formation of a new body that regulates whaling (maybe a commission that is of a similar nature as NAMMCO, but with a global character). While cetaceans enjoy a special standing under the UN Convention on the Law of the Sea and 'conservation of marine mammals and in the case of cetaceans shall in particular work through the appropriate international organizations for their conservation, management and study',[80] it is nowhere stipulated that the IWC is to constitute the only body in this regard. Jefferies, for example, proposes the adoption of a marine mammal agreement under the United Nations.[81]

Consequently, in order to avoid a dissolution of the IWC through potentially more and more states leaving, mere reliance on science appears not to be sufficient, given the diverging interpretations of scientific findings. I argue therefore that it appears consequently imperative to insert a different element – cultural communication – in order to provide new impetus for IWC members to find a solution to the standstill within the organisation.

Currently, and to the best of my knowledge, no international environmental organisation has a committee or a similar sub-body in place that discusses cultural dimensions of conservation and sustainable use. While the Convention on International Trade in Endangered Species of Wild Fauna and Flora (CITES) has a working group on livelihoods in place, this is not a permanent body and the Standing Committee, and ultimately the Conference of the Parties, can rather quickly opt to disband this body. Moreover, it depends on the chair of the working group as to how efficiently and ultimately successfully it can operate. Therefore, in order to prevent an implosion of the IWC and to try to find ways to overcome the cultural divide, in this chapter, I propose the establishment of a (sub-)Committee on Cultures and Whales (CCW) that is set up as a permanent body under the IWC to discuss cultural dimensions of whales and whaling – at least as long as no solution to the current situation within the IWC has been found.

Proposing a (Sub-)CCW

The underlying idea of the proposed (Sub-)CCW is that parties to the whaling Convention have the opportunity to discuss different elements pertaining to the conservation and sustainable use of whales beyond biological parameters. Instead, questions and issues surrounding moral aspects of human interaction with the environment are discussed that allows parties to gain an insight into why a certain position on cetacean conservation and utilisation prevails within a given society. This is to say that the CCW aims to delve deeper into the positions of states that are communicated throughout the different deliberations and in the often diverging gists of interventions of parties to the ICRW.

In order to prevent an agenda-driven mandate of the CCW (the persuasion of parties with opposing views), the committee should in principle not be open to both state and civil society observers, in order for IWC members to be able to find compromise in light of diverging views. The main objective of the CCW is therefore to build bridges between parties, to enable them to further explain and outline their views, to find common ground and ultimately to find a solution to the deadlock concerning commercial whaling and whale preservation. As the annual (or chair's) reports show, this deadlock has been identified time and again by different IWC members, but thus far, no way to break this deadlock has been found, especially with regard to the RMS.

The ultimate goal, the foundation of the mandate, is therefore to break the dead-lock that has taken hold of the Commission since the failed adoption of the RMS. The CCW's overall approach would be to establish human beings as an inherent part of the ecosystems, which have up to this point been the realm of biologists. The CCW therefore aims to establish 'a human ecology which incorporates both people's ecological activities and their understanding of the world'.[82] In order to achieve this, the mandate should further contain the following goals:

The CCW aims to enable parties

- To provide a platform to exchange views on whales and whaling, based on cul-tural and historic developments,
- To find a common ground pertaining to the importance and ways of cetacean conservation and utilisation in light of cultural and historic developments,
- To discuss the potential utilisation of cetaceans as a food source, and
- To discuss realistic and feasible alternatives to the potential lethal use of ceta-ceans, especially concerning developing countries with or without a whaling history.

In order to reduce costs, the CCW should not be a mandatory body and should not make decisions based on votes. Instead, the CCW should be inherently consensus-driven, including the option of not finding a consensus. This allows the committee to work without pressure and to allow itself to maintain this 'work in progress' as long as a compromise, a consensus, is found. The CCW then reports to the biannual meeting of the IWC as to its progress and with possible ways to find solutions to the deadlock. While the finding of solutions is not necessarily within the mandate of the CCW, it nevertheless provides observations to the plenary that are taken into account when decisions are made. The CCW therefore serves as a platform to generate and disseminate knowledge on whales and whaling within the cultures of IWC members and, ultimately, to find compromise amidst vast ranges of cultural diversity amongst IWC members.

While the CCW should not be open to observers, it should not be shielded from outside expertise either. This means that in order to further elaborate on the cultural role of whales and whaling within a given society, CCW members can propose and invite credible social sciences experts who are able to provide their views and evi-dence for the way cetaceans and whaling are culturally perceived. These views can then be taken into consideration in order to propose ways forward to the plenary. To exemplify this with regard to the standing of NGOs, an example from Norway shows that Greenpeace has never managed to establish itself very well within Nor-wegian Society because it has underestimated the overall state-friendliness of the Norwegian people while having failed to include local community perspectives. As regards the latter, local communities in Norway are by-and-large shaped by the powerful nature that surrounds them, thus placing the human within nature with dif-ferent resources that can be exploited – including whales. Why one species should be more protected than others is not a narrative that can prominently be found. As a consequence, anti-whaling campaigns have never fallen on an equally receptive ground as in, say, Germany or the UK.[83] Similar views can be expressed in the CCW that provide deep insight into the cultural grounds for a country's position.

While the proposed name of the committee is indeed CCW, it is merely advisory and does not have any decision-making powers. It therefore does not stand on equal footing as the SC, the primary purpose of which is to provide the Commission with scientific advice based on an analysis of scientific data. In a sense, the CCW is of a similar character since it provides social sciences advice to the Commission. However, in order to avoid potential competency discussions, it should be made clear from the very beginning that such a committee is permanent, but solely advisory, with a clear mandate to break the deadlock beyond considerations of biological parameters. This point should be emphasised bearing in mind the failed attempts to establish a Rural Communities Committee under CITES, which was rejected over matters of legal standing and the fact that conservation and not rural communities are within the ambit of CITES.

In this sense, the CCW can also be established as a sub-committee to the SC, both indicating that it does not have the same standing within the Commission, but at the same time indicating that social science considerations are necessary. It is furthermore clear that it is whales and whaling that are discussed, not issues of food security, participation or other more normative issues, which many argue the IWC has no mandate for.

How Likely Is the CCW's Establishment and What Could Its Possible Effects Be?

The establishment of the CCW – at least at present – is not very likely as this could potentially be considered a tool to lift the moratorium. It must therefore be clear from the very outset that this is not the Committee's purpose, but rather that it serves to prevent an implosion of the IWC. As mentioned, an agenda-driven Committee is not what is aimed for, but rather for it to be an avenue for communication. In addition, the inertia by which international organisations such as the IWC function and, presumably, because of the fact that anti-whaling countries are content with the direction the IWC has taken make it very difficult for the Commission to undergo substantive change. If the establishment of the CCW were to be proposed by one or more IWC members, the first hurdle would therefore be to 'sell' the idea to other Commission members and to emphasise its benefits for the IWC's functioning. The main 'selling point' is that cetaceans are neither 'sacred' nor 'profane' – following Durkheim's distinction – and that both views carry their own moral charging that has its own legitimisation in certain cultures. At the same time, it is clear that in a culture where cetaceans are indeed 'profane', this 'profanity' is recognised as standing on equal footing as in cultures where cetaceans enjoy 'sacredness'. In other words, the CCW's main task and its main goal is to achieve a discourse on 'equality in difference' in the perception of cetaceans and their sustainable use. If this can be accomplished, the IWC could take a major leap forward in overcoming its deadlock and to develop new ways and methods to avoid other parties leaving.

As indicated above, one of the main challenges is that anti-whaling nations could view this as a new method that aims to lift the moratorium. This, however, should not be the goal, but rather that both sides of the aisle finally start to understand each other *culturally* in order to develop new pathways of dialogue. The CCW is therefore a tool to improve communication which itself is necessary to overcome the deadlock, rather than a tool to justify or advocate a certain stance on whaling within the Commission. Contrary to studies that were carried out before that aim to culturally justify whaling,[84] the CCW is therefore to bring history and social anthropology into the discourse on whales and whaling. This also means that whaling countries start to better understand the stringent anti-whaling stance of whaling opponents, which are often being accused of being unscientific and purely emotional. And vice versa, whaling opponents get the opportunity to understand why whaling advocates are clinging to their position despite the majority of the IWC members constantly overruling them.

Unfortunately, the IWC is confronted with severe financial challenges, which has caused IWC68 in 2022 to tighten the budget for the Secretariat's staff. A proposal for the establishment of a new (sub-)committee would inevitably lead to financial concerns since any new activity requires a certain budget. Realistically, however, the main work of the CCW would take place in the intersessional period without the requirement to meet in person. Online meetings and online presentations could still serve the purpose of gaining insight into the cultural elements of cetacean conservation and utilisation.

This notwithstanding, the idea should not be dismissed just because it is rather unrealistic. Instead, I consider it worthwhile pursuing despite the challenges it might face, simply because in order to improve the functioning of the IWC, a new approach is necessary. Of course, this does not mean that if the CCW is not established the IWC will implode – this would indeed be exaggerated. But the potential of the CCW to enable IWC members to overcome, or at least discuss, deeply rooted differences pertaining to whales and whaling beyond issues of science and the (lethal) utilisation of whales. To this end, the proposal in this chapter should serve as a new idea that could help strengthening the IWC in the future and to overcome the 'sacred' and the 'profane' within the Commission.

Conclusion

In this chapter, I considered the cultural dimensions on the discourses pertaining to whales and whaling. As shown, a variety of discourses on the conservation and utilisation of biodiversity exist, which inevitably also have implications for international environmental governance. Within the IWC, however, it appears as if a rather binary discursive environment exists: that of the 'sacred' and the 'profane' whale, as per the definition by Emile Durkheim. As such, the discourse on the 'sacred' whale considers whaling only permissible in the context of ASW but refuses any commercial aspect of whaling. Over the course of the several decades, particularly Western whaling states have changed their position on whaling and have become adamant anti-whaling advocates.

While this has occurred on the international level and therefore within the IWC, it is also reflected in the common discourse on whaling in respective societies. The 'sacred' whale has become part and parcel of discourses surrounding whaling, which is exemplified by anti-whaling TV shows such as *Whale Wars* or 'Save the Whale' campaigns paired with an anthropomorphic view on whales advocated in children's magazines such as *Mickey Mouse.* Indeed, it is 'normal' to consider whales 'sacred' and to dismiss any views that are different.

The 'profane' whale, however, is essentially the underlying discourse of the IWC. From its very outset, it was what the Commission was pursuing in so far as whales were considered a resource. As was shown throughout this chapter, however, discrepancies over the way whales were to be used and protected at the same time started to emerge already rather early on. Over time, and especially during and after the 1970s, international discourse replaced this with the 'sacred' whale, which finally led to the adoption of the 'moratorium' in 1982 – carried by the majority of IWC members up to this day.

Some states such as Japan, however, have maintained their 'profane' view on whales but were consistently voted down. I argue throughout this chapter that science cannot resolve this matter and has failed to be applied in a way that sheds light on the question of whether or not a whale is 'sacred' or 'profane'. Instead, the answer to this question should be sought in the cultural perception of whales within IWC member states: what role do they play within a society? How are they perceived by the general public? Do they stand for something else than merely being a resource (or not)? The survey that was carried out for the purposes of this chapter indicates that depending on the country, whales and whaling carry fundamentally different views, which may have led, or at least contributed, to Japan's withdrawal from the Convention. In order to make it more unlikely in the future that other countries follow suit and to overcome the deadlock within the Commission, this chapter therefore proposes the establishment of a (sub-) CCW, which aims to establish a cultural dialogue on whales and whaling within the Commission. While it appears rather unlikely that such a committee will be established, the idea is to serve as a new perspective on the way the IWC could improve its functioning.

Species trigger emotions within people. And cetaceans in particular probably stand at the top of the list of species frequently referred to as 'charismatic megafauna'. But to merely pin attitudes towards whales and whaling on emotions is arguably oversimplified. Instead, the cultural environment of a person having developed these emotions should be considered. In that sense, 'traditional' scientific approaches may not be sufficient, but a social sciences perspective can indeed be helpful. It can be hoped that the IWC will overcome its differences in the future, but, as this chapter has hopefully shown, a new, culture-based perspective is necessary to be successful.

Notes

1 Nikolas Sellheim, 'Quotas, cultures, and tensions. Recent schedule amendments for aboriginal subsistence whaling under the international convention for the regulation of whaling', (2018), *Curr Dev Arc L* 6, 4.
2 See for example Edward B. Tylor, *Primitive culture* (J. Murray 1871); Ralph Linton, *The study of man: An introduction* (D. Appleton-Century 1936); Marvin Harris, *Culture, people, nature: An introduction to general anthropology* (Thomas Y. Crowell 1975); or Roger M Keesing, *Cultural anthropology: A contemporary perspective* (Holt, Rinehart and Winston 1981).
3 Footnote omitted; Aliaksandr Birukou, Enrico Blanzieri, Paolo Giorgini and Fausto Giunchiglia, 'A formal definition of culture', in Katia Sycara, Michele Gelfand and Allison Abbe (eds.), *Models for intercultural collaboration and negotiation* (Springer 2013) 1, 4.
4 Michael Heazle, *Scientific uncertainty and the politics of whaling* (UWP 2006).
5 A critical view on the precautionary principle is provided in Cass R. Sunstein, 'Beyond the precautionary principle', (2002), John M. Olin Program in Law and Economics Working Paper No. 149.
6 Cf. the different whaling programmes conducted by the Institute for Cetacean Research (ICR), <https://www.icrwhale.org/Research.html> accessed 14 April 2023.
7 IWC, The RMP — a detailed account, <https://iwc.int/rmp2> accessed 14 April 2023.
8 IWC, RMS, <https://iwc.int/index.php?cID=581&cType=html&zenario_sk_return= zenario__content/panels/content_types/item//html//item//en//html_1> accessed 17 April 2023.
9 Nikolas Sellheim, 'The stereotype is dead? No! Long live the stereotype — A review of Avatar: The way of water', (2023), *The Polar Connection* <https://polarconnection.org/ review-avatar-way-of-water-stereotype/> accessed 17 April 2023.
10 Arne Kalland, *Unveiling the whale. Discourses on whales and whaling* (Berghahn Books 2009) 28.
11 Andrew T. Guzman, *How international law works. A rational choice theory* (OUP 2008).
12 Kirkpatrick Dorsey, *Whales and nations: Diplomacy on the high seas* (UWP 2016).
13 See also Guzman, *note 13*, 170.
14 Robert L. Friedheim, 'Negotiating in the IWC environment', in Robert L Friedheim (ed.), *Toward a sustainable whaling regime* (UWP 2001) 200, 207.
15 E.g. Johnathan R. Strand and John P. Tuman, 'Foreign aid and voting behavior in an international organization: The case of Japan and the international whaling commission', (2012), *Foreign Policy Analysis* 8(4), 409.
16 *Sellheim Environmental*, 'The tabled resolution on food security', (2022), *TCLD* 1(4), 22.
17 Emile Durkheim, *The elementary forms of religious life* (George Allen & Unwin Ltd 1915).
18 Ibid., 40.
19 Jason Brownlee, Tarek Masoud and Andres Reynolds, *The arab spring: Pathways of repression and reform* (OUP 2015).
20 Stephen R. Kellert, 'American attitudes toward and knowledge of animals: An update', in M.W. Fox and L.D. Mickley (eds.), *Advances in animal welfare science 1984/85* (HSUS 1984).
21 Stephen R. Kellert, 'Attitudes, knowledge, and behaviour towards wildlife among the industrial superpowers: United States, Japan, and Germany', (1993), *JSI* 49, 177, 179.
22 E.g. John Braithwaite and Valerie Braithwaite, 'Attitudes towards animal suffering: An exploratory study', (1982), *IJSAP* 3(1), 42; Wolfgang Schulz, 'Attitudes toward wildlife in West Germany', in Daniel J. Decker and Gary R. Goff (eds.), *Valuing wildlife: Economic and social perspectives* (Westview 1987).
23 Linda Kalof, 'The multi-layered discourses of animal concern', in Helen Addams and John Proops (eds.), *Social discourse and environmental policy. An application of Q methodology* (Edward Elgar 2000), 174.

24 Deborah Kelemen, Sarah A. Brown and Lizette Pizza, 'Don't bug me!: The role of names, functions, and feelings in shaping children's and adults' conservation attitudes about unappealing species', (2023), *JEP* 87.
25 Batholomeo Jerome Chinyele and Noel Biseko Lwoga, 'Participation in decision making regarding the conservation of heritage resources and conservation attitudes in Kilwa Kisiwani, Tanzania', (2019), *JSHMSD*.
26 Leonie K. Fischer, Lena Neuenkamp, Jussi Lampinen ... and Valentin H. Klaus, 'Public attitudes toward biodiversity-friendly greenspace management in Europe', (2019), CL.
27 Roger Doménico Randimbiharinirina, Torsten Richter, Brigitte M. Raharivololona, Jonah H. Ratsimbazafy and Dominik Schüßler, 'To tell a different story: Unexpected diversity in local attitudes towards Endangered Aye-ayes *Daubentonia madagascariensis* offers new opportunities for conservation', (2020), *P&N*.
28 Nikolas Sellheim, 'The legal question of morality: Seal hunting and the European moral standard', (2016), SLS 25(2), 141.
29 Hiroyuki Watanabe, *Japan's whaling. The politics of culture in historical perspective* (Trans Pac Press 2009); Jun Morikawa, *Whaling in Japan: Power, politics and diplomacy* (C Hurst & Co Publishers Ltd 2009).
30 Nikolas Sellheim, *The seal hunt. Cultures, economies and legal regimes* (Brill 2018).
31 Herluf Sigvaldsson, 'The international whaling commission: The transition from a 'Whaling Club' to a 'Preservation Club', (1996), *Cooperation and Conflict* 31(3), 311.
32 ICRW, last preambular paragraph.
33 IWC, *Third Report of the Commission* (Office of the Commission 1952) 3.
34 IWC, *Third Report of the Commission* (Office of the Commission 1953) 6.
35 IWC, *Eighth Report of the Commission* (Office of the Commission 1957) 17.
36 IWC, *Ninth Report of the Commission* (Office of the Commission 1958) 14.
37 Sellheim, *Seal hunt.*
38 IWC, *Eleventh Report of the Commission* (Office of the Commission 1960).
39 Rachel Carson, *Silent spring* (Houghton Mifflin 1962).
40 Eliza Griswold, How 'Silent Spring' ignited the Environmental Movement. *The New York Times Magazine.* <https://www.nytimes.com/2012/09/23/magazine/how-silent-spring-ignited-the-environmental-movement.html> accessed 19 April 2023.
41 Charlotte Epstein, *The power of words in international relations. Birth of an anti-whaling movement* (MIT Press 2008) 100.
42 Robert Poole, *Earthrise. How man first saw the Earth* (YUP 2008).
43 EPA, EPA history: First Earth Day. <https://www.epa.gov/history/epa-history-earth-day> accessed 19 April 2023.
44 IWC, *Twenty-second Report of the Commission* (Office of the Commission 1972) 23.
45 Ibid., 27.
46 IWC, *Twenty-third Report of the Commission* (Office of the Commission 1973) 6.
47 United Nations, *Report of the United Nations conference on the human environment* (United Nations 1972) Recommendation 33.
48 IWC, *Twenty-third Report of the Commission* (Office of the Commission 1973) 38.
49 IWC, *Twenty-fourth Report of the Commission* (Office of the Commission 1974) 25.
50 IWC, *Thirty-third Report of the International Whaling Commission* (The International Whaling Commission 1983) 21.
51 Dorsey, *Whales and Nations,* 267.
52 Ibid., 274–277.
53 Justin McCurry, 'Japan split on whale hunts, poll shows', *The Guardian* (30 November 2012) <https://www.theguardian.com/environment/2012/nov/30/japan-whale-hunts-survey> accessed 19 April 2023.
54 AFP, 'Sixty percent of Japanese support whale hunt' *Phys.org* (22 April 2014) <https://phys.org/news/2014-04-sixty-percent-japanese-whale.html> accessed 19 April 2023.
55 Arne Kalland and Brian Moeran, *Japanese whaling. End of an era?* (Routledge 2011); Hiroyuki Watanabe, *Japan's whaling. The politics of culture in historical perspective* (Trans Pacific Press 2009); Nobuhiro Kishigami, Hisashi Hamaguchi and James M.

Savelle (eds), *Anthropological studies of whaling* (National Museum of Ethnology 2013).

56 Mayumi Itoh, *The Japanese culture of mourning whales: Whale graves and memorial monuments in Japan* (Palgrave Macmillan 2008).

57 See also Jefferies and Stock in this volume.

58 James E Candow, *Of men and seals. A history of the newfoundland seal hunt* (Environment Canada 1989) 116–125.

59 Peter J. Stoett, *The international politics of whaling* (UBC Press 1997) 91.

60 Kalland, *Unveiling the Whale* 99.

61 The disc can be listened to on YouTube <https://www.youtube.com/watch?v=-6K7Vno3bU0> accessed 19 April 2023.

62 Charles Flowers, 'Between the harpoon and the whale' *The New York Times* (New York, 24 August 1975).

63 IWC, *Twenty-third Report of the Commission* (Office of the Commission 1973).

64 IWC, *Twenty-fourth Report of the Commission* (Office of the Commission 1974) 24.

65 IWC, *Twenty-sixth Report of the Commission* (Office of the Commission 1976) 25.

66 Ibid., 26.

67 E.g. Science News, 'IWC sets commercial whaling moratorium', *Science News* 122(5); Philip Shabecoff, 'Commission votes to ban hunting of whales', *The New York Times* (New York, 24 July 1982).

68 United Nations, 'Growth in United Nations Membership' <https://www.un.org/en/about-us/growth-in-un-membership#1980s> accessed 12 October 2022.

69 Kalland, *Unveiling.*

70 William Aron, 'Science and the IWC', in Robert Friedheim (ed.), *Toward a sustainable whaling regime* (UWP 2001) 105, 117.

71 George F. McHendry, '*Whale wars* and the axiomatization of image events on the public screen', (2012), *EnCom* 6(2), 139.

72 Christopher Robé, 'The convergence of eco-activism, neoliberalism, and reality TV in whale wars', (2015), *JFV* 67(3–4), 94, 101.

73 Ibid.

74 Sellheim, 'The stereotype'.

75 IWC, *Chair's report of the intersessional meeting on the future of IWC* (The International Whaling Commission 2008).

76 Bundesministerium für Ernährung und Landwirtschaft, 'Whale conservation' <https://www.bmel.de/EN/topics/fisheries/marine-protection/whale-conservation.html> accessed 23 October 2023.

77 Deutscher Bundestag, 'Ja zu Beitrittsverhandlungen' <https://www.bundestag.de/webarchiv/textarchiv/2010/29383731_kw16_de_eu_island-201580> accessed 23 October 2023.

78 For a summary of the resolution, see Sellheim Environmental, 'The tabled resolution on food security', (2022), *TCLD* 1(4), 22. <https://sellheimenvironmental.org/the-conservation-livelihoods-digest/> accessed 4 January 2023.

79 Heazle, *Scientific uncertainty.*

80 United Nations Convention on the Law of the Sea 1982, Art 65.

81 Cameron SG Jefferies, *Marine mammal conservation and the law of the sea* (OUP 2015).

82 Kay Milton, *Environmentalism and cultural theory. Exploring the role of anthropology in environmental discourse* (Routledge 1996) 59.

83 Kristin Strømsnes, Per Selle and Gunnar Grendstad, 'Environmentalism between state and local community: Why Greenpeace has failed in Norway', (2009), *Environmental Policy* 18(3), 391.

84 Hiroyuki Watanabe, *Japan's whaling. The politics of culture in historical perspective* (Trans Pacific Press 2009).

Part III
Perspectives

10 Commercial and Institutional Impacts of Japan's Withdrawal from the Whaling Convention – A Commentary

Gavin Carter

Japan's decision to leave the International Whaling Commission (IWC) (or more precisely its decision to withdraw from the International Convention for the Regulation of Whaling) raised concerns about the implications for the management of global whale populations. Japan's move also raised broader questions about how global commerce might be impacted in the future and, specifically, the extent to which this decision may lead to changes in the global management of wildlife and ocean resources.

Prior to leaving the IWC, Japan overcame two considerable challenges to its whaling program. It defeated Sea Shepherd in the U.S. courts, forcing the activist group to end its dangerous physical sabotage of whaling in the Antarctic,[1] and Japan reconstituted its research program after the International Court of Justice (ICJ) deemed that it was not sufficiently purposed for science.[2] The victory over Sea Shepherd established that law and order on the High Seas is not beyond the reach of the legal system. For Japan, this removed a significant impediment to future commercial whaling in the Antarctic. The second action largely nullified attempts by Australia and New Zealand to force Japan to end its activity in what they regarded as their 'backyard.'

Nevertheless, the economics of research whaling were always challenging and the reconstituted program took fewer whales. Ultimately, reduced sales of whale products would be expected to lead to lower revenues and larger government subsidies. Production costs in research harvests are higher than for commercial hunts because of the various added elements involved in undertaking a complex scientific program. The harvesting process is also slower and uses less efficient methodology.

With these cost pressures, further investment in research whaling could only be justified if there was a realistic likelihood that it would lead to commercial harvesting. While the science showed that sustainable harvests could be implemented, thirty years of negotiations at the IWC made clear that commercial quotas would not be sanctioned.

Japan had considered leaving the IWC over many years (at least since the reform process collapsed at IWC62 in Agadir, Morocco in 2010), carefully weighing up the pros and cons while trying to persuade a majority of members to tackle the instrument's dysfunctionality. As these reform efforts failed to make headway, a

DOI: 10.4324/9781003250814-13

consensus finally emerged among Japan's government agencies that progress was not possible. Withdrawal from the ICRW eventually followed.

From a practical perspective, whaling businesses lost nothing from the withdrawal that had not already been taken away. As a non-member of the IWC, Japan's whalers are no longer allowed to harvest whales in the Antarctic – but this is simply preventing them from doing what they already could not do on a commercial basis. The targeted minke whale population in the Antarctic remains abundant but no other nation has an interest in exploiting this resource or has permission to do so if it is a member of the IWC. Aside from the relatively narrow whaling interests, therefore, Japan's decision has had little direct economic effect.

The Future of the IWC

Looking at wider impacts, it is clear that the IWC itself is now in a more precarious position. Its income has fallen substantially without Japan's financial contributions.[3] Perhaps even more significantly, Japan's withdrawal has highlighted how the IWC no longer is geared to meeting its mission to regulate the whaling industry. In January 2021, IWC Chair Andrej Bibič wrote that:

> The IWC is as important and relevant today as it was in 1946 because it has managed to evolve in response to an ever-changing environment. […] [S]hip strikes, ocean noise, marine debris, entanglement, chemical pollution and of course, climate change […] were largely non-existent when the IWC was formed.[4]

Of course, it is more accurate to say that these problems were not generally recognized (in the case of climate change) or were not considered to be an area of concern for the organization when it was established. This suggests that 'mission creep' has developed markedly at the IWC. Whether or not this is a good thing depends on one's perspective and it will be interesting to see how far the IWC can evolve. The clash over whale welfare and the development of offshore wind farms creates a particular new dilemma for environmental advocacy.

The IWC is now seventy-five years old and is sometimes referred to as the first global conservation treaty. The only whale harvests that it regulates today are those conducted by indigenous populations, placing it in an invidious position as the arbiter of their quotas and harvesting requirements. Under an antiquated and offensive process, the IWC requires indigenous communities to justify their annual dietary whale requirements. Inevitably, this raises complaints about cultural imperialism, colonialism and racism.[5]

Some have predicted that the IWC could collapse. If so, this would not prompt an environmental catastrophe. Whaling today is relatively limited and demand for whale products is not expected to increase. The whale has been saved and harvesting is not the biggest threat to whale populations. Indigenous communities would surely celebrate being freed from the IWC's shackles.

Sovereignty

What lessons can be learned from Japan's decision to leave the IWC and could the institution's troubles be a portent for the future? One obvious fear is that as international environmental instruments become more and more prescriptive, they undermine national sovereignty. The IWC regulates whaling within the waters of each member state, allowing non-whaling members to dictate harvests to the few remaining whaling nations. Since both Norway and Iceland are not bound by the zero quota rules, this means that only indigenous whalers in the USA, Russian Federation, Greenland and St. Vincent and the Grenadines are subject to the demands of other IWC member nations.

By withdrawing from the ICRW, Japan's new commercial whaling, carried out within its 200 nautical mile Exclusive Economic Zone (EEZ), is not subject to the same external oversight. But by carefully regulating catches to ensure that they are sustainable, Japan is demonstrating how the IWC is obsolete in a world with modern management systems. This further explains the need for the IWC to evolve.

The Convention on International Trade in Endangered Species of Wild Fauna and Flora (CITES) is another international instrument that some observers argue has deviated from its original purpose and become too politicized. Some complain that its international trade regulations are encroaching on sovereign domestic laws. Might some African countries, angered by limits on exporting ivory, follow Japan's IWC approach and withdraw from CITES? Alternatively, fearing withdrawals, might CITES reform itself and return to its core functions?[6]

The IWC impact may also be felt by the International Seabed Authority (ISA) which has jurisdiction under UN Convention on the Law of the Sea (UNCLOS) over the recovery of precious metals from under the oceans. Unless alternative solutions are discovered, the desire to reduce reliance on fossil fuels as expressed in the Paris Climate Accord will increase demand for these metals, such as through mandates for electric vehicles in the transportation sector. It is also notable that it proved difficult to finalize the new Biodiversity Beyond National Jurisdiction (BBNJ) agreement – the text was finally agreed in March 2023, much later than had been anticipated.

International Regulation and Science

These issues point to the difficult question of the role of science in shaping treaty obligations that relate to the natural environment. The IWC Scientific Committee was populated with experts who had genuine scientific credentials but who, in most cases, also had opinions that went beyond their expertise, such as on the ethics of hunting whales.

The IWC and CITES demonstrate how decision-making in international conventions can become corrupted when science and politics are mixed. While this observation is not unique to international wildlife institutions (disagreements within the World Health Organization about the origins of Covid-19 come to mind), it has produced scientifically unjustifiable protections for abundant minke whales and abundant mako sharks. This begs the questions, what standing does science have in the IWC when abundant minke whales are prohibited from being harvested?[7] What standing does science have in CITES when it lists in its Appendices a species of shark with a global population of around 20 million? Will the global community countenance similar prohibitions in other resource management organizations in the future, such as the ISA or BBNJ?

Despite the modern call to defer to what 'the science' demands, scientific enquiry is an ever-moving phenomenon. Ultimately, this means that decision-makers in international meetings should make knowledge-based decisions while resisting the temptation to invoke 'science' to support political positions.

Nonprofit Sector

While nonprofits are generally not considered to be businesses, their tax status tends to conceal their substantial fundraising and spending. Many of the organizations historically associated with lobbying at the IWC earn hundreds of millions of dollars each year and together employ thousands of lawyers, scientists, lobbyists and marketing specialists.[8]

When Japan left the IWC and thereby forfeited its Antarctic whaling, it also took away one of the most iconic campaign slogans of the last fifty years – the effort to 'save the whale.' This has forced NGOs to choose whether to refocus their campaign against Japan's domestic whaling or shift their attention to other issues. So far, they have taken the latter course.

Many of these non-governmental organizations receive substantial grants from a plethora of philanthropic foundations, themselves established by billionaires to shelter their business wealth and provide a vehicle through which to build an environmental legacy. So far, it seems that the nonprofit sector has chosen to focus its efforts on prohibiting the exploitation of ocean resources in 30% of the oceans by 2030 (the so-called '30x30' target) and 50% by 2050. If this move is only partially successful, it could have a devastating impact on different sectors of the global economy.

Most obviously, it could decimate fisheries, leading to a general recalibration of food production as consumers seek alternatives to traditional seafood. The 30x30 movement that is already backed by many developed nations like the United Kingdom and part of the recently adopted Kunming-Montreal Global Biodiversity Framework[9] under the Convention on Biological Diversity (CBD) would also impact energy extraction, shipping, pharmaceuticals, cabling, tourism and other sectors both directly and indirectly.

Business Impacts

Most corporations and businesses spend very little time pondering these questions. However, business transactions in the future may be particularly susceptible to external challenges on the grounds of environmental responsibility. And it is the multilateral institutions that define what this means.

A recent example arose when advocates were able to persuade a group of airlines to cease transporting lawful wildlife products. The immediate consideration for the airlines was whether continuing to ship the products was worth the negative publicity. They quickly decided that it wasn't. However, such considerations are rarely straight forward. Two former Secretary Generals of CITES criticized the airlines, pointing out how their embargoes undermined international cooperation, did not actually help the species in question and threatened the livelihoods of people in developing countries.[10]

In general, the business community has paid little attention to the growing impact of United Nations agencies or international treaties and agreements on wildlife and the oceans. Instead, they tend to maintain a reliance on legal and public affairs professionals to lobby their national officials. For users of natural resources, such a reliance on past practices is outdated and places at risk the broader commercial environment in which they operate.

Fishers are at particular risk from the economic effects of environmental regulations. They are also at a considerable disadvantage in presenting their viewpoints globally because the industry is characterized by many small businesses, most of which do not have the resources – in terms of money, manpower and time – to understand and counter the regulatory risks they face. Collective institutions such as fisheries associations are better placed to act but are burdened by a need to satisfy a broad membership constituency which, in turn, tends to dilute actions that can be taken.

In Japan, government officials work much more closely with domestic businesses than their counterparts in most Western countries. Japan's decision to leave the IWC was driven not by the fishers but by Fisheries Agency officials, who were unable to deliver commercial quotas. In other words, the dysfunctionality of an international body directly impacted the ability of the government to perform its purpose. Even so, it took more than twenty years of negotiation and cajoling at the IWC before withdrawal was finally sanctioned.

Western countries follow a different culture, with the relationship between government officials and businesses based more on consultation. Since at least the 1980s, and markedly after the establishment of the Intergovernmental Panel on Climate Change (1988), environmental groups have maintained strong relationships with officials in Western governments, gaining equal or greater access than businesses and, more recently, collaborating as partners.

Thus negotiations at the United Nations to establish an agreement on BBNJ took place with minimal business input, even though its impact on world commerce would be huge. Environmental interests participated in the discussions and lobbied governments and regional groups intensely, hosting workshops and other collaborative events.

As the impact of international treaties grows, corporations and businesses would be wise to take a more considered and holistic view about the global management of ocean resources. Their thinking should take into account reputation, their bottom line and ethical aspects of resource use. These questions go beyond normal public relations, legal obligations and corporate responsibilities. One challenge for businesses is for them to work out how can they do this when they have limited expertise with these complex issues. Another is to overcome a natural reluctance to look beyond their immediate fiduciary purpose of making money for their shareholders.

Conclusions

Since it left the IWC, Japan has not been subject to the same level of organized global criticism for undertaking whaling that it experienced during the preceding forty years. Government officials in Tokyo braced themselves for strong international criticism when Japan announced its withdrawal from the IWC but, in the event, the reaction was muted. In part, this was because the move had little impact on other nations. Historically vocal anti-whaling nations like Australia and New Zealand may even feel they benefited because Japan ended its research program in the Antarctic, which they regard as their 'backyard.'

While the direct impact of withdrawal on commerce was minimal in a global context, it would be a mistake for businesses to believe that the example of Japan's departure from the IWC provides evidence of endemic frailty in the system of international environmental regulation. It is true that Japan left the IWC after the institution became dysfunctional but the pressure to withdraw did not come from business. Therefore, it does not follow that Western businesses would have any success in resisting international environmental regulations in the future by persuading their national leaders to withdraw from treaties.

Equally, Japan's withdrawal from the IWC should not be interpreted as a rejection of multinationalism. This was a labored and reluctant move, taken after many years of implementing alternative strategies, and it was adopted only after all other options had been exhausted. In so far as it sets a precedent, it is that Japan is wedded to multinationalism except in the most extraordinary of circumstances.

Japan's withdrawal from the IWC will have little impact on the global system of treaties that governs environmental issues. In this regard, Brexit (the UK's withdrawal from the European Union) and the United States leaving and then rejoining the Paris Climate Accord may be thought of as being coincidental rather than the beginning of the fracturing of multinational decision-making.

By establishing a 'tipping point,' Japan's withdrawal from the ICRW nevertheless begins to answer the academic question of how far international treaties can go in forming obligations to protect species and the broader natural ecosystem. When science breaks down as the regulatory guide and the integrity of the instrument is subsumed by political considerations, criticality is reached.

Politics and science are sure to converge again in international environmental governance, with CITES, the ISA and BBNJ offering perhaps the best examples. However, with shared benefits from international cooperation, and plenty to be gained in terms of responsible environmental stewardship, it is to be hoped that these institutions will be able to learn from the IWC experience.

Notes

1 AFP, 'Sea Shepherd agrees $2.55m payment to Japanese whalers for injunction breach', *The Guardian* (London, 10 June 2015) <https://www.theguardian.com/environment/2015/jun/10/sea-shepherd-payment-japanese-whalers-breaching-injunction> accessed 21 March 2023.

2 See Malgosia Fitzmaurice and Dai Tamada (eds.), *Whaling in the Antarctic. Significance and implications of the ICJ judgment* (Brill 2016).

3 For information on funding of the IWC, please see IWC, 'Funding' <https://iwc.int/commission/iwcfinancing> accessed 23 April 2023.

4 IWC, 'Marking a milestone: 75 years of the IWC', AB/JAC/32857, 19 January 2021.

5 To demonstrate the long-standing debate surrounding this issue, see Nancy C Doubleday, 'Aboriginal subsistence whaling: The right of inuit to hunt whales and implications for international environmental law', (1989), 17 *DJILP* 373.

6 See Sellheim's CITES chapter in this volume.

7 On science in the IWC, see Michael Heazle, *Scientific uncertainty and the politics of whaling* (CCI Press 2006).

8 See Fernand Vincent, 'NGOs, social movements, external funding and dependency', (2006), *Development* 49, 22.

9 Convention on Biological Diversity, 'Kunming-Montreal global biodiversity framework', CBD/COP/15/L25   <https://prod.drupal.www.infra.cbd.int/sites/default/files/2022-12/221222-CBD-PressRelease-COP15-Final.pdf?_gl=1*sxmosl*_ga*NTQwMTg3MDIuMTY4MjM1ODA4Nw..*_ga_7S1TPRE7F5*MTY4MjM1ODA4Ny4xLjAuMTY4MjM1ODExNC4wLjAuMA..> accessed 23 April 2023.

10 CITES, 'CITES Secretary-General calls on airlines to reconsider boycotts of wildlife shipments', 4 May 2001 <https://cites.org/eng/news/pr/2001/010504_LH.shtml> accessed 23 April 2023.

11 Whales on the Rise, the IWC Demise and Global Environmental Diplomacy

An Epilogue to the Whaling Wars – A Commentary

José Truda Palazzo, Jr.

A Treaty from Another Planet and an Evolving War of Attrition

The year 1946 is literally a world away from the present. When the International Convention for the Regulation of Whaling (ICRW) was signed in Washington on 2nd December, after World War II did away with the short-lived first whaling treaties signed in 1931 and 1937,[1] and despite the successive depletion of whaling grounds already a well-established fact ever since Japan began bay whaling in the 17th century and the Basques wiped out the Northeastern Atlantic right whales from 700 onwards, the main concern of whaling nations wasn't necessary sustainability (despite the nice words in the Convention's preamble), but the urgent need to share established catches among the fleets of different flags.

At that time, cetacean research was merely a butcher's job. Whaling was *the* source of scientific knowledge on whale species and populations, and literally nothing was known about their ecology and ethology other than the fact that many species had distinct feeding and breeding grounds where the catcher fleets operated, sometimes with lucky researchers being allowed a berth on board to measure, weigh and make observations on the hunt and the carcasses it yielded. It was based on this rather limited, mostly descriptive science that the International Whaling Commission (IWC), the decision-making body established by the 1946 treaty and charged with managing whaling worldwide, would make its 'management' decisions. The IWC Scientific Committee, under the noxious influence of the 'Maximum Sustainable Yield' (MSY) concept used (wrongly enough) in fisheries, would gather data from those patchy funeral observations and catch reports, throw it all in theoretical models and issue what was considered the 'best possible scientific advice' to the Commission Plenary for the establishment of catch quotas. The sheer lack of scientifically sound information on the diverse species targeted would lead to the adoption of bizarre, un-scientific concepts such as the 'Blue Whale Unit' (BWU), based on relative size and weight, to determine catches of species which had nothing to do with blue whales. To make things worse, the IWC, over decades, made a point of regularly ignoring even that sketchy advice and often set catch quotas well above those advised by the Scientific Committee.

DOI: 10.4324/9781003250814-14

We all know the result of the IWC's historic irresponsibility as a management body. One after another, whale species and populations were severely depleted, many to the brink of extinction, with only the comparatively small and uneconomic minke whales spared until the second half of the 20th century. Not managing catches in the sense of wise resource use control, but merely reacting to the sequential demise of its *raison d'être*, the IWC tacitly accepted the role of villain that millions of people around the world would associate with it in the growing whale conservation movement of the 1970s and 1980s, and by not fulfilling its promise as a serious management organization, bought its ticket to oblivion (although the journey would take decades and is still ongoing).

The Inception of a Bipolar World at the IWC and the 'Forgotten' Stakeholder Countries

As the political posturing on whaling raged on, two realities 'in the field' continued to evolve towards a world where whaling would no longer be deemed necessary or acceptable. First, the continued decline in the market for whale products and in their associated value, requiring hefty government subsidies and/or promotion to keep whaling viable in Japan and other countries still engaged in whaling. Second, the growing importance of non-extractive uses of whales, both as generators of jobs and revenue for coastal communities – especially in developing countries – and as providers of vital ecosystem services, according to mounting scientific evidence. Regardless of the IWC and its members' antics, these two real-world factors will likely ensure that whaling as a commercial enterprise will disappear during our lifetime, with only minuscule (but still controversial) so-called subsistence hunts remaining in isolated pockets of national waters.

Perhaps the greatest irony of all is that whaling could have subsisted for much longer if only Japan and the whaling bloc had acknowledged the existence of a third force in the IWC, a group of definitely pro-conservation countries but with a different take on the issues at hand. Latin American countries comprise the bulk of this third force, the so-called Buenos Aires Group, which peeled off from the whaling camp over time, entirely by domestic political movements of their civil societies, and a few other 'non-aligned' countries such as South Africa. Since 2005 the Buenos Aires Group has maintained a strong support for the commercial whaling moratorium, but, during several negotiating processes held at the IWC, trying to break the bipolar deadlock, the bloc signaled its equally strong desire to negotiate in good faith with the whaling countries, based on a few core principles which were very dear to the group: (a) recognition of non-lethal uses of whales, especially whale watching, as a legitimate management option for IWC member states, to be taken into account and respected at any management decision involving shared whale populations; (b) establishment of a South Atlantic Whale Sanctuary focused on basin-wide cooperation for non-lethal research and conservation management of whales; and (c) the withdrawal of far-ranging whaling fleets from the Southern Ocean (Whale Sanctuary), hence from the Southern Hemisphere, where no range States were whaling anymore. These core principles were repeatedly stated not only in joint Opening Statements at Plenary meetings and regular speeches in meeting debates but also in many private intersessional conversations which I witnessed first-hand over the years.

Although large Northern NGOs did try to encroach on the Buenos Aires Group sovereign positions occasionally by touting their own agendas, those positions were the result of true sovereign decisions taken by national governments in close consultation with their own civil society stakeholders. But this important fact seemed to slip from the view of both warring sides. The whalers considered the Buenos Aires Group and associated African partners simply as puppets of the Northern conservation movement and did not make any real attempt to negotiate taking their stated interests into account. On the other hand, the 'like-minded' bloc, with the notable exception of Australia and occasionally New Zealand, took the pro-conservation position of the group for granted and made no actual effort to include their core positions and needs in the full negotiating process.

This diplomatic neglect was worsened by two episodes during the years of attempts to break the Commission's deadlock. First, the constitution of a supposedly 'middle-ground' bloc led by Sweden in the 1990s, self-appointed as the 'Open-Minded Group', which did its best to portray the Buenos Aires Group as 'radical' and tried frequently to push for an abandonment of their core principles as part of a 'negotiation' with the whalers (the 'open-mindedness' of the bloc lost all credibility when the Swedish Chair of the Commission cast a decisive vote, in a Special Meeting in 2002, to allow Iceland to rejoin the IWC with a clearly illegal objection to the moratorium, and for Iceland to illegally vote on its own behalf before being formally admitted back!).[2] Then, more recently, by the Chilean Chair Disaster – a regrettable episode in which a Chilean Ambassador who was Chair of the Commission, after another many rounds of fruitless negotiations between warring sides, decided in 2010 to push the other Buenos Aires Group members to accept giving Japan a commercial whaling quota in its own jurisdictional waters without its whalers having to leave the Southern Ocean Whale Sanctuary. Throughout the whole succession of dramas at intersessional negotiations which lasted many years, Japan refused to cede a single inch in its attempt to maintain a whaling stronghold in the Antarctic. Its allies Norway and Iceland, although having no interest whatsoever in going whaling in the Southern Hemisphere, never missed a chance to publicly oppose, despise and mock the core principles of the Buenos Aires Group, thus contributing to an atmosphere of distrust and entrenchment of positions. Alas, it is common knowledge that the many Caribbean, Pacific and African countries allied with Japan at the IWC (some would say controlled by it) also regularly played this sorry role of obstruction and mockery, to no one's benefit.

The result of all this attrition is well-known. In the 67th IWC Plenary meeting held in Brazil, Japan's definitive attempt to reopen commercial whaling failed, and the Buenos Aires Group countries, with overwhelming support from the like-minded bloc, passed instead the Florianópolis Declaration proposed originally by Brazil and the Latin countries based on a draft suggested by the Brazilian Humpback Whale Institute, and which not only reaffirms the maintenance of the moratorium but also instructs the Secretariat and IWC subsidiary bodies to redirect resources to non-lethal management and conservation of whales. The Japanese delegation announced that very day its withdrawal from the IWC, thus putting an end to what I used to call the Komatsu Paradox. Dr Masayuki Komatsu, Alternate Commissioner for Japan for many years, was one of the sharpest minds in the IWC and a formidable opponent of the anti-whaling bloc. Throughout his many colorful interviews to the international media, I've got the impression that he considered the debates at the Commission as a mockery: whaling countries would simply ignore the moratorium and get on with their whaling under diverse guises; anti-whaling countries would proselytize for a couple of weeks per year against the whaling villains and would get back to their usual deforestation, pollution and overfishing; and wealthy Northern NGOs would have their limelight moments with the media and continue fundraising on the issue with no resolution in sight, thus implying that not changing the *status quo* would be good for everyone. Now the Komatsu Paradox is no longer valid. So what's next for whalers, whales and the world?

Japan's Departure as a Good Thing for Global Whale Management – and for Japan

I am certain to irritate many friends in the environmental movement with this commentary piece, and perhaps rightly so – after all our common goal has always been to end whaling, but I have oftentimes differed from the tactics and directions some colleagues from the North have tried to set in stone. Now, many are trying to portray Japan's withdrawal from the 1946 treaty as a threat and are pushing 'like-minded' countries to criticize it frequently and try to bring Japan back. I disagree on both moves.

First, it seems to me that Japan's decision, after all a logical one, has achieved much of what the Latin countries and their civil societies have been aiming for years, namely the withdrawal of whaling fleets from the Southern Hemisphere. Further, by confining Japan's whaling to its jurisdictional waters, the killing of whales by Japanese fleets has nearly halved. Lastly, there's some hope that over time Japan will stop its practice of recruiting and haranguing small islands and other developing nations to sustain an opposition to whale conservation measures at the IWC (a practice long acknowledged, including by none other than the above-mentioned Dr Komatsu in public interviews).[3]

Restructuring its whaling policy based on domestic catches is good for Japan. This policy has for a large part shielded it from the severe global criticism it received over decades, and whatever protest appears now is only a shadow of the former anti-Japan campaigning. Further, it can hopefully now manage its whaling policy in a more rational way than just reacting badly to foreign pressure. Reducing government subsidies has already begun, and as I write it appears clear that there's no longer enough of a market to sustain commercial whaling by itself.[4]

In the end, Japanese whaling will apparently die an honorable death, sunk by market forces. This will have two main beneficial results. First and foremost, it will relieve Japan from the (real or perceived) role of lead villain in global marine conservation, allowing for its sincere collaboration in developing adequate mechanisms to conserve marine biodiversity including improved international agreements and helping to raise public awareness of much more serious marine conservation issues, including systematic overfishing by Chinese and European fleets.[5] Second, it will most likely do away with the 'whaling bloc' mentality at other international agreements, whereby Japan, Iceland and Norway, with the help of a fluctuating cohort of followers, systematically boycott marine conservation initiatives. A post-whaling Japan has the potential to emerge as a major, positive marine conservation force in the international diplomatic scene. True, there will always be questions about other domestic issues such as the Taiji dolphin hunts,[6] but my bet is that it won't be enough to influence Japanese foreign policy anymore.

Remaining Issues on Whaling and Whales: A Portrait of Modern Non-management

Whaling is finished as a major international environmental matter. What remains of it will largely be confined to byzantine discussions about the models used to calculate aboriginal subsistence whaling quotas – another anachronistic legacy of the 1946 Convention which Japan has legitimately pointed out as hypocritical if compared to its own coastal whaling requests. Denmark, for instance, is allowed to kill hundreds of whales in Greenland, supposedly for its aboriginals *sensu* ICRW, but the meat ends up illegally – and quite openly – in supermarkets way beyond Greenlandic boroughs as demonstrated in several reports,[7] which the European Union – supposedly a major adversary of whaling – continues to ignore. Alaskan tribes have long abandoned the strict 'subsistence' requirement of proof of the Convention to spouse a convolute justification of 'cultural' needs while renewing their quota requests at the IWC. How long these localized, often intrinsically cruel hunts will last – and how dedicatedly some of the largest and wealthiest 'animal welfare' NGOs will continue to shove it under the rug while striving to criticize Japan – is up for debate, but it seems quite clear that they won't become a major issue for international diplomacy to deal with.

Undoubtedly there are still global whale management issues that require the joint effort of the international community. Humankind continues to kill thousands of whales a year through ship strikes and entanglement in fishing gear, including recent incidents in the Antarctic itself where overfishing of krill and overlapping of krill fishing fleets with vital whale feeding grounds emerge as serious concerns. Plastic pollution is taking a heavy toll on odontocete cetaceans, as inferred from a growing number of necropsied carcasses around the world, attesting to slow deaths caused by choking of the animals' digestive system by an array of plastic debris. And the overarching catastrophe of rapid climate change is already disrupting food sources and may reshape coastlines and other features at large whale breeding grounds.

Addressing these issues would require an integrated view of whale management which seems to be utterly lacking from all current international instruments (including the IWC). And it would also require the acknowledgment of scientific evidence regarding the vital importance of whales as keystone species, acting as nutrient and carbon conveyors across ocean provinces and systems. Further, it would entail a definitive convergence between marine spatial planning and species conservation, which is still utterly lacking in all current treaties and agreements, not the least because of turf wars regarding 'jurisdictions' of these instruments. In this regard, the ludicrous lobby by the UN's Food and Agriculture Organization (FAO) and Regional Fisheries Management Organizations (RFMOs) at international fora, including the recently concluded UN negotiations of a treaty to conserve biodiverse beyond national jurisdictions, confirmed a tendency by the fisheries organizations to try and prevent any interference with their disastrous *laissez-faire* activities in the high seas. Exempting large-scale fisheries from international scrutiny and due restrictions is a bad omen for any attempt to achieve the conservation and any actually sustainable use of marine biodiversity – including whales. But it can happen – especially if the whaling controversies at the IWC stop being part of the picture.

Is a Future without the IWC so Bad? Or Does the World Need a Zombie Whaling Treaty?

Under these circumstances, and with Japan's departure signalling, a further reason to de-mobilize the 'armies' of both pro- and anti-whaling camps at the IWC, what kind of future does the Commission have, if any? Will it survive the end of commercial whaling as an activity of international relevance and become a conservation organization, as dreamt by many activists and some governments over the last few years, even as its founding treaty and history of mistakes reek with the smell of dead whales past, or will it simply fade away while other, more modern international agreements and agencies take upon them the tasks of cetacean management in a more ecosystem-oriented context? The first two to five years of a post-pandemic reality will likely tell us the answer.

As I write, it seems unlikely that the IWC will rise to the task of becoming a post-whaling, integrated management framework, able to deal with the multi-layered tasks related to whale management and conservation. Despite brave attempts by its former Secretary to make the Commission emerge from the quagmire of whaling-centered bickering, through a reinforcement of cooperation attempts with other organizations and stimulation of an internal dialog about restructuring the Commission to deal with modern challenges, there doesn't seem to be enough high-level political will from any of the major players to move in that direction. Neither will the original treaty be renegotiated. Anyone wishing for a plenipotentiary conference to amend the ICRW in order for it to reflect a contemporaneous view of whale management is only tossing coins in a wishing well.

Truth be told, Japan is right on yet another count: the 1946 International Convention for the Regulation of Whaling *is a whaling treaty*, and no matter how much we try to legitimately interpret its provisions in an evolving sense, as all legal interpretation should be done, there are insurmountable flaws in the letter of the treaty that gives ground to all the difficulties in running the IWC in the 21st century. As much as I would like it to be otherwise, my impression is that the IWC will not survive the end of Japanese commercial whaling, which seems to be approaching fast. While that doesn't happen, the IWC will still function, if only manned by a skeleton crew of pro- and anti-whaling countries. Why? Because it is imperative for Japan to maintain a certain number of its pro-whaling supporters there should the anti-whaling countries blink and allow for an eventual ¾ majority to be attained to reverse the commercial whaling moratorium; likewise, non-whaling countries need to ensure that Japan and/or its satellite countries do not reach that magic number, thus bringing whaling to the high seas once again. Everything else is in truth secondary, despite the continuous attempts by some governments and NGOs to continue making progress on non-lethal management issues, and by cetacean researchers to keep the Scientific Committee going as their own little realm in the sky, with little regard for the fact that the IWC is an international treaty body, not a scientific congress.

Most likely, at some point in the future, the IWC will be replaced in the international scene by whale-related management action by other treaties, including the Convention for the Conservation of Antarctic Marine Living Resources (CCAMLR), the Convention on the Conservation of Migratory Species of Wild Animals (CMS) and, alas, the new treaty on biodiversity beyond national jurisdiction. Residual whaling (by any excuse or denomination) inside national 200 nautical mile Exclusive Economic Zones (EEZs) will be dealt with domestically, or at the utmost bilaterally or among countries sharing common whale populations. There will no longer be a need for an IWC and, as its many widows weep, it will quietly and honorably – at last – follow the path of all the other dinosaurs into oblivion, at the same time as a diverse array of birds fly away with its DNA on their ancestry, hopefully fostering a new generation of marine conservation projects and agreements much more successful, and more consensual, than that one instrument signed in Washington on one foggy winter day in 1946.

Notes

1 See Jefferies and Stock in this volume.
2 See Couzens in this volume.
3 See, for instance, Jonathan Watts, 'Japan admits buying allies on whaling', *The Guardian* (London, 19 July 2001) <https://www.theguardian.com/world/2001/jul/19/japan.whaling> accessed 23 April 2023.
4 See Julian Ryall, 'Is Japan's whaling industry going under?' *DW* (Bonn, 28 July 2022) <https://www.dw.com/en/is-japans-whaling-industry-going-under-as-demand-sinks/a-62626007> accessed 23 April 2023.
5 Jonathan G. Odom, 'Europe's double standard for China's overfishing', (2020), EJIL: Talk! <https://papers.ssrn.com/sol3/papers.cfm?abstract_id=3577628> accessed 23 April 2023.
6 David McNeill, 'Taiji: Japan's Dolphin cull and the clash of cultures' 5 *APJ* 1, <https://apjjf.org/-David-McNeill/2306/article.html> accessed 23 April 2023.
7 See for instance Whale and Dolphin Conservation, 'Greenland's expanding commercial whaling', <https://uk.whales.org/wp-content/uploads/2018/08/Greenland-expanding-commercial-whaling.pdf> accessed 14 May 2023.

Index

For Product Safety Concerns and Information please contact our EU
representative GPSR@taylorandfrancis.com
Taylor & Francis Verlag GmbH, Kaufingerstraße 24, 80331 München, Germany

www.ingramcontent.com/pod-product-compliance
Lightning Source LLC
Chambersburg PA
CBHW060256220326
41598CB00027B/4125